AF317043

Hydrodynamics of Wave-Vegetation Interactions

Advances in Coastal and Ocean Engineering

ISSN: 1793-0731

Series Editor: Philip L-F Liu *(Cornell University, USA)*

Published

Vol. 5   Advances in Coastal and Ocean Engineering
           edited by Philip L-F Liu

Vol. 6   Advances in Coastal and Ocean Engineering
           edited by Philip L-F Liu

Vol. 7   Advances in Coastal and Ocean Engineering
           edited by Philip L-F Liu

Vol. 8   Advances in Coastal and Ocean Engineering:
           Interaction of Strong Turbulence with Free Surfaces
           edited by M Brocchini, D H Peregrine and Philip L-F Liu

Vol. 9   PIV and Water Waves
           edited by John Grue, Philip L-F Liu and Geir K Pedersen

Vol. 10   Advanced Numerical Models for Simulating Tsunami
           Waves and Runup
           edited by Philip L-F Liu, Harry Yeh and Costas Synolakis

Vol. 11   Advances in Numerical Simulation of Nonlinear Water Waves
           edited by Qingwei Ma

Vol. 12   A Guide to Modeling Coastal Morphology
           by Dano Roelvink and Ad Reniers

Vol. 13   Numerical Modeling of Water Waves in Coastal and
           Ocean Engineering
           by Pablo Higuera, Jinghua Wang, Jie Hu and Zhengtong Yang

Vol. 14   Hydrodynamics of Wave-Vegetation Interactions
           edited by V Sriram, Thorsten Stoesser, Shiqiang Yan and K Murali

To view the complete list of the published volumes in the series, please visit
http://www.worldscientific.com/series/acoe

ADVANCES IN
COASTAL AND OCEAN ENGINEERING
VOLUME 14

Hydrodynamics of Wave–Vegetation Interactions

Editors

V. Sriram
Indian Institute of Technology Madras, India

Thorsten Stoesser
University College London, UK

Shiqiang Yan
City, University of London, UK

K. Murali
Indian Institute of Technology Madras, India

World Scientific

NEW JERSEY · LONDON · SINGAPORE · BEIJING · SHANGHAI · HONG KONG · TAIPEI · CHENNAI · TOKYO

Published by

World Scientific Publishing Co. Pte. Ltd.

5 Toh Tuck Link, Singapore 596224

USA office: 27 Warren Street, Suite 401-402, Hackensack, NJ 07601

UK office: 57 Shelton Street, Covent Garden, London WC2H 9HE

British Library Cataloguing-in-Publication Data

A catalogue record for this book is available from the British Library.

Advances in Coastal and Ocean Engineering — Vol. 14
HYDRODYNAMICS OF WAVE-VEGETATION INTERACTIONS

ISBN 978-981-12-8413-7 (hardcover)
ISBN 978-981-12-8414-4 (ebook for institutions)
ISBN 978-981-12-8415-1 (ebook for individuals)

For any available supplementary material, please visit
https://www.worldscienti ic.com/worldscibooks/10.1142/13619#t=suppl

https://doi.org/10.1142/9789811284144_fmatter

Preface

The hydrodynamics of wave vegetation interactions is a topic of increasing interest to the coastal engineering community over the past decades. Its importance has come into focus after recent natural coastal disasters causing severe damages to the environment, infrastructure and humans. There are a large number of research works carried out to understand the flow physics, design considerations and new dimension of soft engineering in the form of planting of coastal vegetation, compared to the traditional hard measures adopted to protect the coasts from natural disasters. Particularly, research tools such as physical experiments and numerical modelling have been employed to begin to understand the complex hydrodynamics of wave vegetation interaction and it remains a topic of intense research as a number of knowledge gaps still exist. This edited book contains individual chapters from leading researchers in this field who provide summaries of past work, lay out the challenges and complexities of the subject and provide an outlook of where future resources should be dedicated to in order to advance our understanding of wave vegetation interactions. We believe this book will be useful for early-career researchers as well as stakeholders, decision makers and graduate students who would like to pursue their studies in this field. We would like to thank the authors who accepted our invitation to contribute, namely Dr. Maike Paul, Dr. Maria Maza, Prof. Norio Tanaka, Prof. Dominic Reeves, Dr. Zhihua Xie, Prof. Holger Schuettrumpf, Prof. V. Sundar and all the co-authors of the individual chapters.

Finally, this book was edited through a collobrative project between Indian Institute of Technology Madras, India; University College London, UK and City, University of London, UK through funding from the Scheme for Promotion of Academic and Research Collaboration (SPARC) project of Ministry of Human Resource Development. We would like to thank the funding agency for their support and initating the activities on "Hydrodynamics of Wave-Vegetation Interactions"

V. Sriram
Thorsten Stoesser
Shiqiang Yan
K. Murali

Contents

https://doi.org/10.1142/9789811284144_0001

Chapter 1

Introduction to Wave-Vegetation Interaction

Thorsten Stoesser

University College London,

Department of Civil, Environmental and Geomatic Engineering,

Gower Street, London

t.stoesser@ucl.ac.uk

1. Relevance of Wave-Vegetation Interaction

Without a doubt our climate is changing, having just witnessed air temperatures in London beyond 40 degrees Celsius breaking the record set just three years earlier in 2019 at 38.7 Celsius, and with climate change our ice sheets are melting and ocean water temperatures are increasing. Extreme heat and warm oceans energise hurricanes, typhoons, or cyclones or causes "just" strong winds which in turn bring storm surges and large waves towards our coasts. Rising sea levels makes these storms even more treacherous as waves overtop more easily coastal defences and water gets pushed farther inland resulting in extreme flooding, severe damage to infrastructure, coastlines to be eroded away and people's lives being put at risk. Yet, the population density in coastal areas has increased steadily over the last decades with the consequence that many intertidal zones have come under pressure due to the disappearance or degradation of their ecosystems which provide valuable services in terms of coastal protection, such as near-shore vegetation (e.g. sea grass meadows, salt marshes and mangroves) which works well for the purpose of coastal protection.[1–3] Densely populated coastal areas need resilient solutions that are sustainable, scalable, robust, multi-functional and economically feasible.[4]

What follows is that coastal defense is becoming increasingly dependent on engineered solutions including the development of innovative and cost-effective man-made ecosystems to protect coastlines.[5] Alternatively, natural ecosystems in the form of aquatic vegetation on the foreshore, can be completed with engineered infrastructure (so called "hybrid" approaches) which enhance coastal resilience by providing important storm and coastal flooding protection.[6] Vegetated foreshores, such as submerged sea-grass meadows, salt marshes and emergent mangrove forests, are known for their ability in reducing the strength and impacts of waves by dissipating wave energy and/or reduce the impact of storm surges, respectively and hence can provide sustainable coastal protection. Innovative concepts such as *Building with Nature* provide a basis for coastal protection strategies that are able to follow gradual changes in climate and other environmental conditions, while maintaining flood safety, ecological values and socio-economic functions.[4] However, the success of such strategies rely on tools the development of which need a thorough mechanistic understanding of the functioning of ecosystems in order to design sustainable ecosystem-based protection and restoration solutions for coastal resilience planning efforts. The predictive performance of these tools require a deep knowledge of the interplay between vegetation and wave so that reliable engineering forecasting models can be developed or refined. At the moment large-scale forecasting models appear to over-simplify the extremely complex wave-vegetation interaction simply because this process has not yet been fully understood and which is why wave-vegetation interaction has been subject of research for many years and will be for many more.

2. Types of Wave-Vegetation Interaction

Wave-vegetation interaction is a complex, multi-dimensional and highly-unsteady fluid structure interaction process taking place inside vegetation canopies. In coastal areas wave-vegetation interaction may be categorised into different types according to the vegetation species and its properties, their location near the coast and their height relative to the mean water level. Figure 1 illustrates three of those wave-vegetation types: (a) submerged flexible vegetation such as seagrass or seaweeds, (b) emergent and submerged semi-flexible such as cordgrass or other grasses of the saltmarsh, and (c) emergent rigid such as mangroves. Whilst all three types are found to dissipate wave energy, the exact mechanisms at play are very different. For instance, in type (a) wave-induced currents over submerged vegetation are characterised by the so-called canopy shear layer a region near the interface between the unobstructed free-stream flow above and the obstructed

(a) Submerged flexible (seagrass)

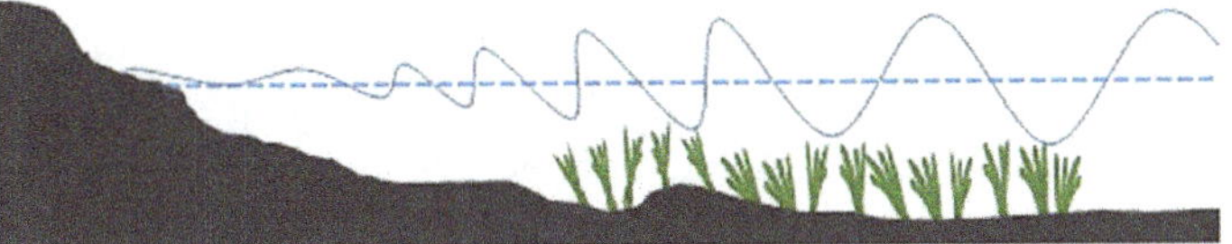

(b) Emergent and submerged semi-flexible (saltmarsh)

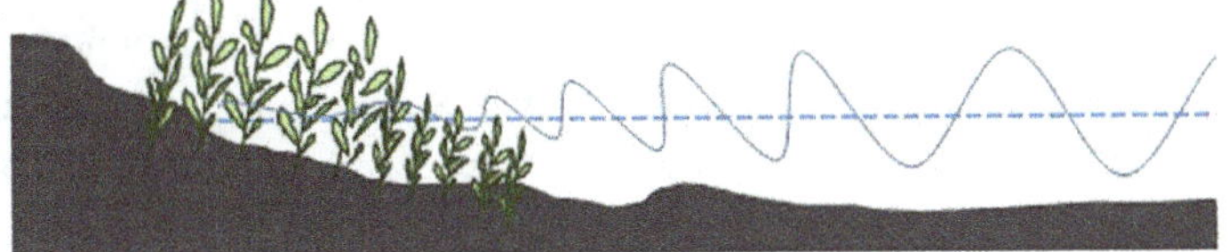

(c) Emergent rigid (mangroves)

Fig. 1.   Different types of foreshore or shoreline vegetation.

in-canopy flow below that interface. Whilst flexible submerged vegetation barely modifies the velocity profiles (in comparison to unobstructed wave-induced currents), submerged rigid vegetation shifts the vertical profile of the wave-induced current upwards and inside the canopy layer stem-generated vortices lead to a local wake behind each stem.[7] Emergent rigid vegetation, type (c), reverses the direction of current-induced waves deep inside the canopy and a strong shear layer develops near the water surface. The wave-vegetation interaction process not only depends on the type of vegetation but also on a number of distinct vegetation and hydrodynamic parameters such as vegetation density, vegetation flexural rigidity, relative submergence or wave height and frequency, etc.

3. Scope and Outline of the Book

In this book, the state of the art in modelling wave-vegetation interaction either by laboratory experiments or numerical modeling/simulation is discussed and recent research carried out by scholars in this field is summarised. Both research approaches face a number of challenges, for example in experiments, one of the research questions is how to scale-down appropriately the large-scale physics of WVI and how to interpret the experimental data. In numerical simulations, the major challenge lies in the extremely wide range of scales ranging from plant scale $(mm - cm)$ turbulence generation due to flow separation to the water depth (several m) in wave-current interaction with emergent vegetation which require adequate numerical resolution, the enormous spatial extent of eco- or engineered-systems and the lack of exact formulations (equations) to compute numerically the complex fluid-structure interaction between waves and flexible vegetation.

Given the urgency of the matter of a warming planet, the book focuses on the specific topic of wave-vegetation interaction in the coastal engineering realm, a complex and fascinating research topic. Nonetheless, it is of critical importance to the engineering community who needs a good understanding of governing physical mechanism of energy dissipation and reliable forecasting tools for the design of ecosystem-based solutions in their quest to protect our coasts from devastating storm surges and extreme waves. A number of established scholars have contributed to this book. In Chapter 2, Dr. Maike Paul, Leibniz University Hannover, introduces and describes the relevant physical properties of vegetation, which vary strongly between species. Her chapter also shows how these properties can be measured and what they depend on. Given the large amount of aquatic species with a wide range of parameters and properties, Harish *et al.* of RWTH Aachen, in Chapter 3 review the effect of coastal vegetation and points out the need for many more studies on hydrodynamics in vegetation to improve the general understanding of processes and the performance and highlight the many challenges experimentalists face. Chapter 4 is dedicated to experimental and numerical modelling of flow-vegetation interaction in the framework of Nature-Based Solutions (NBS) for coastal defence and in this chapter Dr. Maria Maza, University of Cantabria, presents the main physical processes involved in flow-vegetation interaction. Furthermore, she lays out how these processes have been modelled in recent laboratory experiments and numerical models. In Chapter 5, Prof. Norio Tanaka describes the so-called hybrid defence structure, i.e. combining vegetation and hard structures in different arrangements and discusses experiments and field work carried out on the coast of Japan. In Chapter 6, Dr. Thomas van Veelen, Prof. Harshinie Karunarathna, and Prof. Dominic Reeve, Swansea University and University of Twente, provides insights into the role of vegetation flexibility on wave attenuation and evidences via numerical modeling the damping of regular waves by rigid and flexible artificial vegetation. In Chapter 7, Divya Ramesh and Sriram Venkatachalam, IIT Madras, elaborate on the modelling of large-scale domains employing phase-averaged models or phase-resolving models, respectively. Different numerical treatment approaches of coupling fluid with porous structures are discussed and advantages and disadvantages are pointed out. In Chapter 8, Zhihua Xie and Thorsten Stoesser elaborate on the types turbulence structure in wave-vegetation interaction and the advantage of using eddy-resolving numerical simulations in resolving these. They also show a few recent attempts of wave-vegetation interaction simulations using the method if large-eddy simulation. Finally, in Chapter 9, Hari Ram and Sundar Vallam provides a comprehensive review on the wave vegetation interaction studies, highlighting the experimental modelling efforts by the researchers as well as future perspective of the hybrid approaches in wave-vegetation interactions.

This book is useful to the established researcher who can find good summaries and recent advances in the subject but also to students, young researchers, teachers and practicing engineers who are willing and wanting to dig deeper in the challenging yet exciting topic of wave-vegetation interaction.

References

1. S. Temmerman, P. Meire, T. J. Bouma, P. M. J. Herman, T. Ysebaert, and H. J. De Vriend, Ecosystem-based coastal defence in the face of global change, *Nature.* **504**(7478), 79–83 (2013).
2. C. M. Duarte, I. J. Losada, I. E. Hendriks, I. Mazarrasa, and N. Marba, The role of coastal plant communities for climate change mitigation and adaptation, *Nature Clim. Change.* **3**(11), 961–968 (2013).
3. B. Ondiviela, I. J. Losada, J. L. Lara, M. Maza, C. Galván, T. J. Bouma, and J. van Belzen, The role of seagrasses in coastal protection in a changing climate, *Coast. Eng.* **87**, 158–168 (2014).
4. E. van Slobbe, H. J. de Vriend, S. Aarninkhof, K. Lulofs, M. de Vries, and P. Dircke, Building with nature: In search of resilient storm surge protection strategies, *Natural Hazards.* **65**(1), 947–966 (2013).

5. T. J. Bouma, J. van Belzen, T. Balke, Z. Zhu, L. Airoldi, A. J. Blight, A. J. Davies, C. Galvan, S. J. Hawkins, S. P. G. Hoggart, J. L. Lara, I. J. Losada, M. Maza, B. Ondiviela, M. W. Skov, E. M. Strain, R. C. Thompson, S. Yang, B. Zanuttigh, L. Zhang, and P. M. J. Herman, Identifying knowledge gaps hampering application of intertidal habitats in coastal protection: Opportunities & steps to take, *Coast. Eng.* **87**, 147–157 (2014).

6. A. E. Sutton-Grier, K. Wowk, and H. Bamford, Future of our coasts: The potential for natural and hybrid infrastructure to enhance the resilience of our coastal communities, economies and ecosystems, *Envir. Sc. Pol.* **51**, 137–148 (2015).

7. D. Pujol, T. Serra, J. Colomer, and X. Casamitjana, Flow structure in canopy models dominated by progressive waves, *J. Hydr.* **486**, 281–292 (2013).

Chapter 2

Vegetation Traits

Maike Paul

*Leibniz University Hannover, Ludwig Franzius Institute of
Hydraulic, Estuarine and Coastal Engineering,
Nienburger Straße 4, 30167 Hannover
paul@lufi.uni-hannover.de*

Hydrodynamics–vegetation interactions follow the same physical principles, irrespective of vegetation type. However, magnitude of the relevant physical traits of vegetation vary strongly between species, in space and in time. This chapter starts by describing these traits and how they can be measured before it outlines their variations and what these depend on. It concludes with a discussion on how these variations can best be considered when assessing hydrodynamics-vegetation interactions in field, laboratory and numerical applications.

Keywords: Reconfiguration, vegetation zonation, vertical trait distribution, dynamic vegetation traits, sampling method, phenotypic plasticity.

1. Introduction

Coastlines that are exposed to low to medium high hydrodynamic energy can be colonised by vegetation. These plants can usually survive occasional high energy events like storms, but will not be able to sustain under high hydrodynamic forcing over longer time periods. However, up to which level of mean hydrodynamic energy plants can survive strongly depends on the vegetation type and plant species.

Coastal vegetation responds to the forces induced by hydrodynamic energy by reconfiguration. While plastic reconfiguration (e.g. buckling, breaking or tearing) can occur if hydrodynamic forces become too high, elastic reconfiguration (i.e. bending or streamlining) is of interest in the context of hydrodynamic–vegetation interaction. The way coastal vegetation interacts with hydrodynamics follows the same physical principles, irrespective of vegetation type. However, magnitude of the relevant physical traits of vegetation vary strongly between species, in space and in time which needs to be considered when assessing hydrodynamic–vegetation interaction.

2. Types of coastal vegetation

Coastal vegetation can be broadly grouped into four types which each build the foundation of a coastal ecosystem. The borders of the ecosystem are then defined as the boundary where the vegetation type ends and it comprises all living organisms within this area as well as all chemical and physical aspects of this area[1]. While essentially the whole ecosystem will interact with hydrodynamics, the vegetation type will dominate any attenuation processes. The four types (mangroves, saltmarsh, seagrass and macroalgae) will be briefly introduced here with a focus on their general appearance and distribution. For details on individual species, the reader is referred to biology and ecology literature.

2.1. *Mangroves*

Mangroves are terrestrial plants which can grow in water logged soil with salinities from brackish to high saline (9%), but require air exposure of their leaves for most of the tidal cycle. Based on these requirements, they can be found from the supralittoral to the lower eulittoral (Fig. 1) where they colonise soft sediments. They are sensitive to low temperatures and are generally limited to latitudes between 25° N and 25° S, although exceptions can occur in sheltered locations (e.g. Gulf Saint Vincent, South Australia).

Mangroves are trees with a woody rigid trunk that supports a crown with branches and leaves. They are grounded with an extensive root network which partially emerges from the soil for air uptake. Depending on the mangrove species, these aerial roots are developed as prop roots (Fig. 2a) or narrow flanges extending from trunk or branches to the ground or as pneumatophores growing upwards from the underground

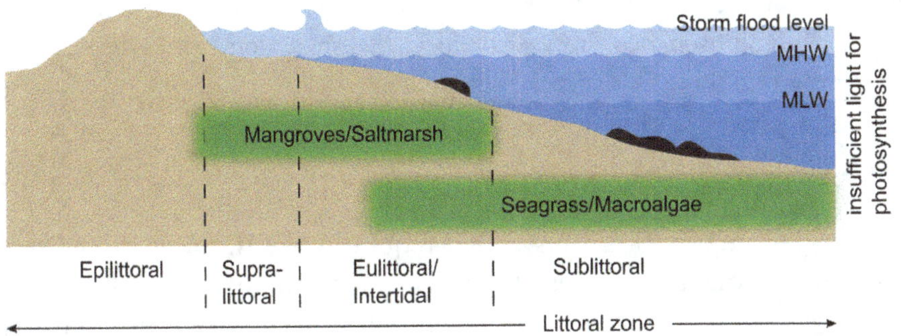

Figure 1. General growth range of coastal vegetation types within the littoral zone. The lower depth limit of the growth range in the sublittoral depends on light availability on the seabed and thus varies between locations. MHW = mean high water, MLW = mean low water.

root network. The latter can be pencil-like (Fig. 2b) or knee-shaped, depending on mangrove species[2].

2.2. *Saltmarsh*

Saltmarsh describes a community of salt-tolerant non-woody plants like grasses and herbs (Fig. 2c). Like mangroves, they are terrestrial plants and thus require air exposure of their leaves for most of the tidal cycle. This defines their lower growth limit in the littoral zone (Fig. 1). Many saltmarsh species can grow in fresh water conditions, but get outcompeted by other species in such locations leading to their upper growth limit in the supralittoral. Saltmarshes can be found on soft sediments across all latitudes with their dominant distribution in the temperate zone.

Within their growth range (Fig. 1) saltmarshes show a general zonation into pioneer zone in the lower eulittoral followed by the low (<mean high water), middle (<mean high water spring) and high marsh (<highest astronomical tide)[3]. Given the different frequency and magnitude of hydrodynamic exposure along this gradient, plant species with different strategies and abilities to withstand the resulting forces colonise the different zones.

In estuaries and other locations where salinity reduces to brackish conditions species composition changes to reeds and rushes. With respect to their interaction with hydrodynamics these ecosystems can be compared with saltmarshes, however, due to their ecological difference they are termed tidal reeds or brackish marshes.

2.3. *Seagrass*

Seagrasses are aquatic plants which require salt water to survive. Some species can tolerate exposure to air for some time of the tidal cycle but most species are restricted to the sublittoral which clearly sets them apart from saltmarshes. Seagrasses grow strip- or petal-like leaves from sheaths protruding from the root/rhizome network. Rhizomes are stem-like organs growing horizontally underground which store nutrients and expand the plant's size by growing new shoots from their tips (Fig. 2d). Seagrasses colonise soft sediment down to the depth where light at the seabed is not sufficient for photosynthesis anymore[4]. The lower depth limit, thus, strongly depends on water clarity while the upper depth limit is driven by turbulence caused by waves.

Seagrasses can be found on all continents apart from Antarctica with number of species declining towards higher latitudes. As a result, many seagrass meadows in temperate regions are monospecific, i.e. comprising of only one seagrass species, while in the tropics meadows with more than 10 different species can be found.

2.4. *Brown macroalgae*

The alternative name seaweed often leads to confusion with seagrass why the term macroalgae is used here. Biologically, macroalgae differ from the other vegetation types described here as they neither have roots nor do they produce flowers and generally have a less complex internal structure. While red and green macroalgae exist, brown macroalgae can be considered most relevant in the context of hydrodynamic–vegetation interaction and resulting engineering applications (e.g. wave and flow attenuation) with kelps making up the largest group of species. Brown macroalgae can occur in all climate regions but are most prominent in cold to temperate regions.

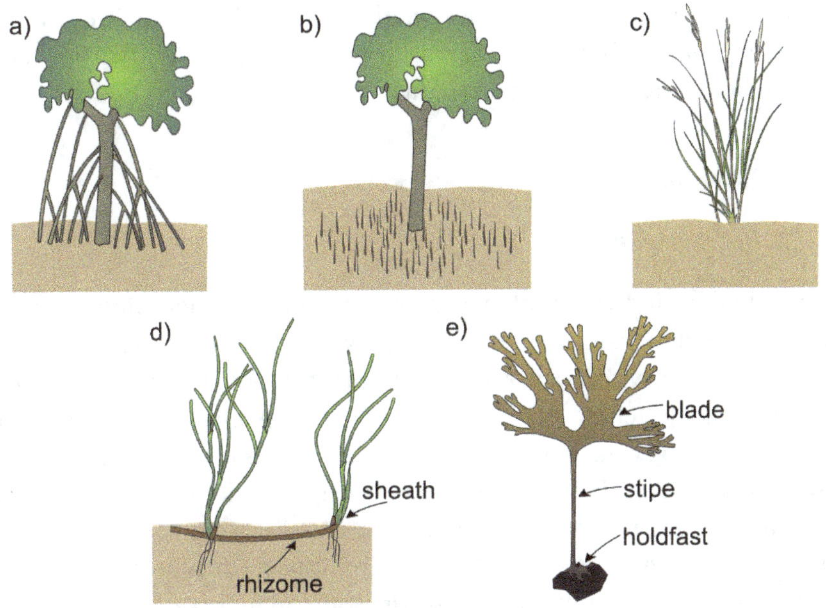

Figure 2. Schematics (not to scale) of different types of coastal vegetation with terrestrial plants in the top row and aquatic vegetation in the bottom row. a) mangrove with prop roots, b) mangrove with pneumatophores, c) saltmarsh, d) seagrass, e) brown macroalgae. Images adapted from the Integration and Application Network (ian.umces.edu/media-library).

Macroalgae require saline water and hard substrate to which they attach themselves to with a holdfast. They colonise rocks, pebbles or large mussel shells from the intertidal down to their light limit somewhere in the sublittoral (Fig. 1). From the holdfast, they extend a stem-like stipe which extends into leaf-like blades of varying size and shape (Fig. 2e). These can contain pea-shaped organs which add to the complexity of their shape.

3. Vegetation traits

The traits relevant for describing vegetation in the context of their interaction with hydrodynamics refer to the aboveground parts of the plants. The belowground biomass comprising of roots and rhizomes becomes relevant when the effect of vegetation on sediment stabilisation is addressed. Since sediment dynamics are closely linked to hydrodynamic forcing, belowground biomass will also be considered here briefly.

3.1. *Aboveground biomass*

3.1.1. *Plant morphology*

Morphology of individual plants can be described by their dimensions, or number and dimensions of their plant parts. The length can easily be obtained *in situ* with a ruler while the plant is being fully extended by hand without stretching it. Saltmarsh and seagrass plants as well as many brown macroalgae usually measure several decimetres while kelp and mangroves can reach several metres in length. Width and thickness of stems and leaves is better measured with a calliper gauge or micrometre screw as they range from sub-millimetre scale (thickness of seagrass leaves) to several centimetres (kelp stipes and mangrove roots). Even for those stems that can be assumed to be circular, it is appropriate to measure both dimensions for the assessment cross sectional area.

As soon as plants grow in stands such as meadows (saltmarsh and seagrass) or forests (mangroves and brown macroalgae) overarching dimensions also become relevant. Shoot density N describes the number of plants per square metre. And apart from mangroves, it is counted within a randomly marked quadrat, usually of 0.5 x 0.5 m, and then scaled up. For very dense meadows the quadrat size can be reduced to 0.25 x 0.25 m although this may reduce the accuracy of the absolute values[5]. Typical values range from $O(1)$ for mangroves and $O(10)$ for kelp to $O(100)$ for saltmarsh and seagrass and can peak in $O(1000)$ for some seagrass species like the intertidal species *Zostera noltei*[5]. Traits that combine the information of individual plant dimensions and shoot density are biomass per square metre and leaf area index. Biomass per square metre is traditionally obtained by harvesting all aboveground biomass from a randomly marked quadrat which is then dried to constant weight over 48 h at 60–80 degrees Celsius. If the vertical distribution of biomass is of interest for the assessment of velocity profiles for instance, the plant material can be placed on a board in its approximate natural position and cut in intervals which are then dried individually[6]. A less destructive method uses image processing techniques to convert photos of plant stands into biomass information, but these technologies are presently limited to non-inundated areas, i.e. mangroves and saltmarsh in the coastal context. Airborne light detection and ranging (LiDAR) and other remote sensing technologies can be used to assess largescale areal biomass distribution[7,8]. Another meadow or canopy trait commonly assessed by remote sensing techniques is leaf area index (*LAI*) defined as the one-sided leaf area per ground area which finds application in a wide range of natural sciences[9]. It has also been found to be a useful variable when describing flow resistance of riparian vegetation[10]. For submerged vegetation or small scale applications *LAI* can also be derived manually by harvesting and measuring all foliage area on a given ground area.

If small scale information on vertical distribution of saltmarsh is required, an *in situ* photo method will be more appropriate. Lateral photographs of the vegetation are taken against a high contrast background which are then converted into binary images[6,11]. From these images the canopy height and frontal area in still conditions can be derived and transfer functions for saltmarsh biomass assessment across the vertical plane have been derived[12]. Canopy height differs from plant length as it describes the plants' heights in their natural position. For flexible vegetation canopy height will be less than plant length as gravity and other forces will lead to some bending. Drag forces due to approaching flow typically lead to a reduction of canopy height due to streamlining which also affects frontal area as relevant trait for hydrodynamics–vegetation interaction. Frontal area describes the plant's or canopy's projected area onto the vertical plane and will thus be affected by a reduction in canopy height. And also the width of a branched plant may reduce due to streamlining under approaching flow as the branches may be compressed towards the plant's centre[13] also affecting frontal area. Canopy height and frontal area are therefore dynamic traits dependent on incident hydrodynamic forcing. Meaningful values can thus only be obtained under the hydrodynamic conditions of interest and will respond dynamically to any fluctuations of the flow. Thus challenges arise for obtaining lateral photographs to quantify canopy height and frontal area without disturbing the flow field.

3.1.2. *Biomechanics*

In addition to hydrodynamic conditions, elastic reconfiguration also depends on plant traits which can be described with methods from structural mechanics and are grouped under the term biomechanics. Following structural mechanics, biomechanics assumes plant parts to be uniform in their shape and isotropic in their internal composition. Considering the complex shape of branching mangroves or brown macroalgae it becomes apparent that uniformity is a strong simplification. The same applies for isotropy given that plants are composed of various cell types with different chemical composition and can contain crevices filled with air or liquid. Biomechanical traits are thus always idealised[14]. However, given the large natural variability of plant traits within a species (see Sec. 4.2.), it is presently considered an adequate approximation.

Flexural rigidity

Considering plants as cantilevers where roots or the holdfast serve as fixed end, beam theory can be applied to obtain the plant's stiffness or flexural rigidity J (Nm^2)[15]. It can be used to describe how a plant resists bending under a point or distributed load and depends on the plant tissue's composition as well as the plant's cross sectional shape. Tissue composition in this context can be described by Young's bending modulus E (Nm^{-2}) which is a mere material trait and independent of the sample's size and shape[15]. While it appears to be attractive for the direct comparison of plant species based on this independence, Young's bending modulus in itself is not sufficient to describe elastic plant reconfiguration to inform hydrodynamic–vegetation interaction since, analogous to cantilevers, elastic reconfiguration of plants also depends on how the plant material is distributed around its neutral axis[15]. This distribution is described by the second moment of area I (m^4) and depends on the cross sectional shape. The most common approximations are circular shapes (I_c) for trunks, stems, stipes and mangrove roots or rectangular cross sections (I_r) for leaves and blades:

$$I_c = (\pi d^4)/64 \tag{1}$$

$$I_r = (bt^3)/12 \tag{2}$$

where d = diameter (m), b = width (m) and t = thickness (m). Triangular, elliptic or hollow circular cross sectional areas are also possible. However, it has been found that such geometrical approximations overestimate cross sectional area for stems of the saltmarsh plant *Elymus sp.* by approx. 10% when compared to area analysis based on microscopical analysis of thin sections[16]. Given the high additional effort and required skills to produce and analyse thin sections, such an overestimation may be acceptable for most practical applications.

For plant parts that are rigid enough to withstand bending by gravitational forcing, flexural rigidity can be derived from 2-, 3- or 4-point bending tests, with results from 3-point bending tests more commonly found in literature. For the latter it has become common practise to use a span width s (m) of 15 times the sample's diameter to reduce the effect of shear stress during measurements. The linear part of the force–deflection curve corresponds to elastic behaviour of the sample and can be used to calculate flexural rigidity:

$$J = (s^3 F)/48D = EI \tag{3}$$

with F denoting the applied force (N) and D the corresponding displacement (m).

Especially leaves or blades are often too flexible to perform such bending tests. Fixation of the sample's ends to prevent slipping due to gravitational force introduces tensile stresses during the test, making computation of flexural rigidity for comparison with other studies difficult[17]. Alternatively, tensile tests have been used to obtain the modulus of elasticity in tension as a measure of biomechanical behaviour. However, tensile stresses are only one component of bending around a neutral axis and the resulting information is thus not sufficient to fully describe the elastic reconfiguration of plants or plant parts. To overcome this problem, Peirce's cantilever test has been modified[18] which uses gravity as the acting force to bend a sample to a predefined angle (Fig. 3). Flexural rigidity can then be computed from the angle and the length of sample required to reach this angle:

$$J = Wl^3 \, [\cos(\theta/2)/(8 \tan(\theta)] \tag{4}$$

With θ = predefined angle, l = bent sample length (m) and W = weight per unit area of the sample (Nm^{-2})

$$W = (w_{wet}g)/(l_{total}b) \tag{5}$$

where w_{wet} = wet weight of the sample (kg), g = gravitational acceleration (9.81 ms^{-2}) and l_{total} = length of sample (m) (Fig. 3). While it has the advantage of assessing bending traits directly, it requires a certain sample length which can be difficult to obtain, e.g. in the case of short leaves.

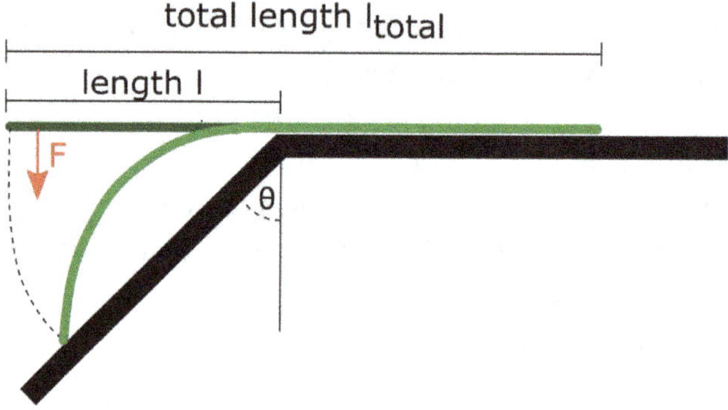

Figure 3. Schematic of the modified Peirce's cantilever test. The force *F* is imposed by gravity to bend the sample to the defined angle *θ*. The length *l* to achieve this angle can be used to derive flexural rigidity *J*. See Ref. 18 for details.

All the above methods have in common that they require samples of uniform width and thickness and in the case of very large plant parts such as kelp blades this requires cutting of appropriate sections for testing. However, cutting also means damaging the plant tissue and it is not known how this affects the sample's biomechanical traits. Equally, desiccation due to air exposure during tests, changes in environmental conditions during storage or damage due to uprooting during plant collection may affect biomechanical traits. Measurements on the seagrass *Posidonia oceanica* suggest that this species' biomechanical traits are insensitive to sampling method and storage time up to 48 h[19]. Biomechanical traits of blade strips from the brown macroalgae *Laminaria digitata*, however, responded to cutting[18] and temperature and salinity of storage conditions[20] albeit these results remained inconclusive. Since the possible effect is presently not known, it is recommended to reduce the time between sample collection and testing as much as possible and store samples in environmental conditions as similar to the conditions at the sample location as possible.

Buoyancy

If plants are submerged, the buoyancy force F_B also contributes to their resistance to bending although this contribution is generally small compared to all other acting forces[14]. For positively buoyant plants or plant parts, F_B always acts vertically upwards and also contributes to the reduction of elastic reconfiguration and restoring an upright posture once hydrodynamic forcing reduces. It depends on the density difference between the plant tissue and the surrounding water and is thus constant for a given plant part:

$$F_B = gV(\rho_p - \rho_w) \tag{6}$$

With V = volume of the plant or plant part (m^3), ρ_p = density of the plant tissue (kgm^{-3}) and ρ_w = density of the surrounding water (kgm^{-3})[21]. The plant's material density can be computed as ratio of its mass and volume and for plant parts with simple geometry, it is possible to obtain the volume from measuring their dimensions. For complex geometries, the displacement method is more suitable where the sample is submerged in water and the increase in water level allows calculation of the displaced volume of water. Since the displaced volume of water equals the volume of the sample's section below the water line, care must be taken to submerge the whole sample. In case of positively buoyant plant parts, this will require a pin or needle to hold the sample down ensuring the tool to be small enough as not to introduce measurement errors by adding submerged volume. Mass is then obtained by weighing the plant parts on a balance with sufficient accuracy, which in the case of individual seagrass leaves may require a sub-milligram scale. However, care must be taken to remove all excess water from the plant's surface prior to weighing. And desiccation during the weighing process can affect the results, especially for small and lightweight parts such as individual seagrass leaves. Consequently, an efficient workflow is advisable in which plant parts are taken out of the water and padded dry with a paper towel directly next to the balance, weighing is done without any delay and samples are returned to water immediately.

3.2. *Belowground biomass*

Belowground biomass has been observed to affect sediment dynamics for seagrass meadows[22,23], saltmarsh surfaces[24,25] and saltmarsh cliffs[26,27], but a full parameterisation of these processes is still lacking. For brown macroalgae belowground biomass is non-existent since they settle on hard substrate. And presently no information is available on the effect mangrove belowground biomass has on sediment dynamics. Ecological descriptions of their structure and function exist (e.g. Ref. 2), but traits relevant for soil stabilisation are rarely described in literature. Given the growing importance of carbon stock assessments for coastal ecosystems[28] and the importance of mangrove belowground biomass in this context[29] more data may become available in the future to improve the assessment of sediment stabilisation and erosion protection by mangroves.

In addition to plant belowground biomass, soil stability also depends on a number of chemical and physical sediment characteristics[30] which are beyond the scope of this chapter. Belowground biomass comprises of roots and rhizomes. The latter are horizontally growing underground stems which plants use to store nutrients and/or for spatial expansion by growing new shoots from their ends. Traits used to describe belowground biomass are root density (in g per m^3) or root length density (in m per m^3) and rhizomes are usually included in these metrics. Several methods to obtain values for these traits exist ranging from simple yet labour intensive separation of soil and roots by rinsing to sophisticated X-ray methods, but no methodological consensus has been reached so far[31,32]. But irrespective of method, it can be necessary to separate a sample in depth layers similar to aboveground biomass prior to processing as root density reduces with sediment depth[33] which has implications for soil stability and sediment dynamics[34,35].

4. Trait variation

The described vegetation traits vary on multiple spatial and temporal scales with differences between species or plant parts (e.g. trunk vs. leaf) being the most obvious reason for variation. But even within the same species, plant or plant parts, trait variations of one order of magnitude can occur. Thus, if values from a single plant are used to assess hydrodynamic–vegetation interactions, results may not be representative for the species and should not be generalised in any case. To account for variations, plant traits need to be described by statistical parameters which are based on sample sizes as large as possible. However, several practical limitations to sample size exist:

- All coastal vegetation types addressed here are protected in many regions of the world. As sampling always means a disturbance of the habitat, it should be kept to a minimum. In some cases permissions are necessary for site access and sample collection and the regulating bodies may limit access duration or the number samples granted;
- Any measurement that cannot be done *in situ* should be done as quickly as possible after sample collection to reduce potential deterioration of the sample. A maximum storage time of 48 h has been established for biomechanical measurements based on the experiences with *Posidonia oceanica*[19] and only a limited number of samples can be processed within this time window;
- Some methods require expensive instrumentation (e.g. computer tomography) or consumables (e.g. if simultaneous chemical analysis is part of the study) and therefore budget may be a limiting factor to sample size.

As a consequence, sample sizes of 10–30 have become common practice, with smaller sample sizes not advisable if statistical comparison of sampled populations is intended.

4.1. *Within plant variation*

On plant part level stem tapering towards the tip has been observed for several saltmarsh species[12,36] and a simultaneous reduction in flexural rigidity has been found[12] highlighting the approximate nature of the uniformity assumption. Such changes along the stem could, however, not be confirmed by a more recent study for *Elymus sp.*[16]. Specimens collected for the two studies varied in size (46 ± 12.3 cm[12] vs. >70 cm[16]) and are likely to have been exposed to different environmental conditions (see Sec. 4.2.). Knowledge on along plant variation of their morphology and biomechanic traits can, however, be relevant for small-scale modelling of hydrodynamics–vegetation interaction as it affects plant bending and thus canopy height and frontal area[37].

Plant tissue changes its chemical composition over time which can affect its biomechanical behaviour. Seagrass leaves have been observed to have higher flexural rigidity in summer than in winter[38] and when they are younger compared to older leaves of the same plant[19] suggesting that they become stiffer during the maturing phase in spring and summer and more flexible again during ageing. It is thus important to sample plant parts of the same growth stage and in the same season when comparing the biomechanical traits of populations from different locations. Seasonality can also affect dimensions on the meadow scale, which is particularly pronounced in temperate regions[39]. Shoot densities of many seagrass and saltmarsh species reach a maximum in summer and a minimum in late winter[5,40] with some species shedding their aboveground biomass completely[41]. Seasonal effects on morphological plant traits (e.g. shoot length, diameter), however, seem to depend on species with some brackish marsh species showing increased values in summer[40] while others and the seagrass *Zostera noltei* have been found to be insensitive to seasonal changes[5,40]. Given this heterogeneity in plant response to seasonal changes careful consideration of input values, validity range and transferability of results is required when hydrodynamics–vegetation interaction is modelled or predicted based on field data of vegetation traits.

4.2. *Spatial variation*

Meadow or forest scale dimensions vary spatially depending on their age, i.e. young plant stands usually have lower shoot density and canopy height compared to well-established ones. In healthy plant stands this effect results in within-stand variations as the outer fringes often comprise of young scattered plants while mature plants cover the stand's centre at high densities. However, spatial variation also depends on environmental conditions. In the intertidal the complex interplay of biological, geomorphological and hydrodynamic processes leads to a heterogeneous pattern of densely or sparsely vegetated and vegetation-free areas[42].

On the scale of individual plants, morphology and biomechanical traits depend on environmental conditions. Interestingly, most of these adaptations are reversible if plants are transplanted to sites with different environmental conditions which is termed acclimation as subcategory of so called phenotypic plasticity[43] which describes an organism's response to the specific conditions it lives in. However, as environmental conditions act in unison, it is difficult to isolate the drivers behind these spatial trait variations within a plant species. Seagrass plants of the genus *Zostera*, for instance, may grow longer leaves in deep water to access more light for photosynthesis in the higher regions of the water column[44]. Shorter leaves in shallower water may, however, also be caused by higher mechanical stresses due to wave motion which are reduced with less exposure of plant tissue to the resulting forces[45,46]. And while seagrasses tend to become more flexible and more resistant to breaking towards colder latitudes both within[38,47] and across species[48], it is uncertain whether this is driven by the lower temperatures, the increased hydrodynamic forcing generally attributed to higher latitudes or other environmental parameters[38].

Hydrodynamic forcing is a strong driver for phenotypic plasticity with almost all aquatic and intertidal plants responding to it following either an avoidance or tolerance strategy by minimising the encountered forces (avoidance) or maximising resistance to resulting damage (tolerance)[49]. The choice of strategy thereby appears to depend on species: The brackish marsh species *Bolboschoenus maritimus*, for instance, was found to produce higher stem diameters at the meadow edge compared to further inside the meadow increasing their resistance to mechanical failure under the higher hydrodynamic forcing on the meadow edge[41]. For the saltmarsh species *Spartina anglica*, however, the opposite trend in stem diameter and flexural rigidity was observed[50], suggesting an avoidance strategy by minimizing the effect of hydrodynamic forcing. A comparison of the sampling locations of the two studies suggests a difference in hydrodynamic forcing with the *Bolboschoenus* site being flow-dominated and the *Spartina* site being wave-dominated. Comparative studies to elucidate how these differences in hydrodynamic forcing affect phenotypic plasticity in marsh species are presently lacking. For intertidal red and brown macroalgae, on the other hand, acceleration forces under wave action have been found to be an important limiting factor for plant size across species[51–53].

In contrast to acclimation, instantaneous elastic reconfiguration (e.g. streamlining, Sec. 3.1.1.) under varying hydrodynamic forcing also leads to trait variation, but does not fall under phenotypic plasticity. Increased wave energy flux, for instance, increases the plants' maximum bending angle under the wave cycle[54] and increased flow velocity will reduce canopy height[55]. These processes also act very locally within plant stands where turbulent vortices can generate a waving motion of the canopy called monami[56].

5. Implications for hydrodynamics-vegetation interactions

As outlined above, plants bend and streamline (i.e. change their frontal area) under hydrodynamic forcing which depends both on biomechanical plant traits and hydrodynamics. The frontal area of vegetation, together with its biomass distribution, in return affects the shape and magnitude of turbulent eddies and the velocity profile and thus the hydrodynamic force acting on the plants (Fig. 4).

On plant scale this leads to a reduction in flow velocity along the streamlined plant height while the velocity profile above the plant remains largely unaffected[57]. In a study with plant surrogates of varying cross sectional shape and flexural rigidity also a pronounced flow reduction at the plant's tip was observed[57] which agrees with observations on the meadow scale. Here an increase in turbulent kinetic energy at the top of the canopy can be observed for a range of hydrodynamic conditions[55] which likely leads to local flow reduction. Turbulent kinetic energy and the size of turbulent eddies generated at the top of the canopy depend on shoot density and thus biomass[55,58].

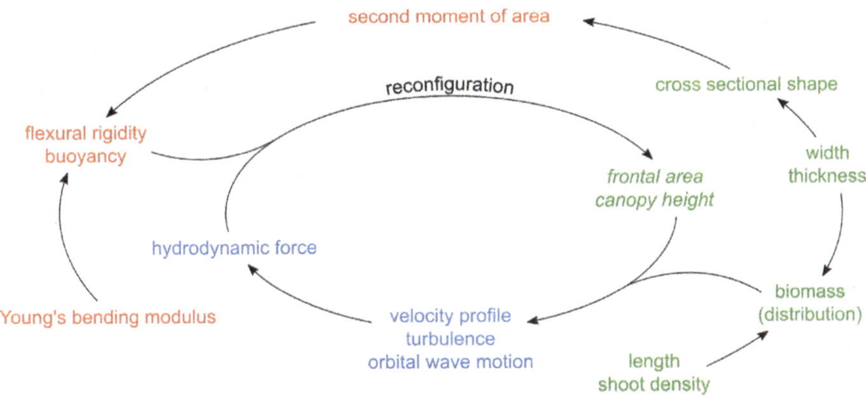

Figure 4. Schematic of hydrodynamics–vegetation interaction with biomechanical (red), morphological (green) and dynamic (green, italics) vegetation traits as well as hydrodynamic descriptors (blue) and reconfiguration as the linking process.

But even for homogeneous meadows with constant shoot density variations can be observed based on the hydrodynamics–vegetation interaction: The front of the meadow is exposed to a given flow velocity which leads to a certain frontal area, the resulting reduced flow velocity exerts less forces further inside the meadow leading to an increase in canopy height and thus frontal area which leads to a stronger reduction in flow velocity and so forth. For flexible surrogates mimicking seagrass this effect has been observed over 2 m of meadow[59] and it is hypothesised that an equilibrium will be reached at a certain meadow length depending on initial flow velocity and the plants' biomechanical traits.

The physical principles hydrodynamics–vegetation interaction is based on are independent of plant species (Fig. 4) and systematic studies with plant surrogates have been used to elucidate the role of individual morphological and biomechanical vegetation traits in this interaction[57,60,61]. To fully understand the implications for the resulting flow field, the effect of spatial variation within a meadow and seasonal variation needs to be assessed in greater detail. Moreover, the importance of natural variability due to tissue age or composition within vegetation traits is not yet fully understood in the context of small scale hydrodynamics–vegetation interactions. For larger scale effects of these interactions like wave and flow attenuation[5,41,62] or plant survival[63] and distribution[64] mean values will often be sufficient, especially as most studies distinguish between species or vegetation types rather than vegetation traits. And while studies have increasingly started to also obtain data on relevant vegetation traits, a full trait-based, species-independent parameterisation is still outstanding.

References

1. F.S. Chapin, P.A. Matson and P.M. Vitousek, *Principles of Terrestrial Ecosystem Ecology*. Springer New York, New York, NY (2011).
2. F. Göltenboth and S. Schoppe, Mangroves. In F. Göltenboth, K.H. Timotius, P.P. Milan, J. Margraf (Eds.), *Ecology of Insular Southeast Asia: The Indonesian Archipelago*, pp. 187–214, Elsevier (2006).
3. C. Chirol, I.D. Haigh, N. Pontee, C.E. Thompson and S.L. Gallop, Parametrizing tidal creek morphology in mature saltmarshes using semi-automated extraction from lidar, *Remote Sensing of Environment* **209**, 291–311 (2018).
4. C.M. Duarte, Seagrass depth limits, *Aquatic Botany* **40**(4), 363–377 (1991).
5. M. Paul and C.L. Amos, Spatial and seasonal variation in wave attenuation over *Zostera noltii*, *Journal of Geophysical Research* **116**, C08019 (2011).
6. U. Neumeier, Quantification of vertical density variations of salt-marsh vegetation, *Estuarine, Coastal and Shelf Science* **63**(4), 489–496 (2005).
7. R.W. Kulawardhana, S.C. Popescu and R.A. Feagin, Fusion of lidar and multi-spectral data to quantify salt marsh carbon stocks, *Remote Sensing of Environment* **154**, 345–357 (2014).
8. R.B. Salum, P.W.M. Souza-Filho, M. Simard, C.A. Silva, M.E. Fernandes, M.F. Cougo, W. do Nascimento and K. Rogers, Improving mangrove above-ground biomass estimates using LiDAR, *Estuarine, Coastal and Shelf Science* **236**, 106585 (2020).
9. G.P. Asner, J.M.O. Scurlock and J. A. Hicke, Global synthesis of leaf area index observations: Implications for ecological and remote sensing studies, *Global Ecology and Biogeography* **12**(3), 191–205 (2003).
10. J. Jalonen, J. Järvelä and J. Aberle, Leaf area index as vegetation density measure for hydraulic analyses, *Journal of Hydraulic Engineering* **139**(5), 461–469 (2013).
11. I. Möller, Quantifying saltmarsh vegetation and its effect on wave height dissipation: Results from a UK East coast saltmarsh, *Estuarine, Coastal and Shelf Science* **69**, 337–351 (2006).
12. F. Rupprecht, I. Möller, B. Evans, T. Spencer and K. Jensen, Biophysical properties of salt marsh canopies — Quantifying plant stem flexibility and above ground biomass, *Coastal Engineering* **100**, 48–57 (2015).

13. J. Aberle and J. Järvelä, Hydrodynamics of vegetated channels. In P.M. Rowiński, A. Radecki-Pawlik (Eds.), *Rivers—physical, fluvial and environmental processes*, pp. 519–541, Springer (2015).

14. V.I. Nikora, Hydrodynamics of aquatic ecosystems: An interface between ecology, biomechanics and environmental fluid mechanics, *River Research and Applications* **26**(4), 367–384 (2010).

15. K. Niklas, Plant biomechanics: An engineering approach to plant form and function. University of Chicago Press, Chicago (1992).

16. J. Liu, S. Kutschke, K. Keimer, V. Kosmalla, D. Schürenkamp, N. Goseberg and M. Böl, Experimental characterisation and three-dimensional modelling of Elymus for the assessment of ecosystem services, *Ecological Engineering* **166**, 106233 (2021).

17. M. Paul, P.-Y.T. Henry and R.E. Thomas, Geometrical and mechanical properties of four species of northern European brown macroalgae, *Coastal Engineering* **84**, 73–80 (2014).

18. P.-Y.T. Henry, Bending properties of a macroalga: Adaptation of Peirce's cantilever test for in situ measurements of Laminaria digitata (Laminariaceae), *American Journal of Botany* **101**(6), 1050–1055 (2014).

19. C.B. de los Santos, B. Vicencio-Rammsy, G. Lepoint, F. Remy, T.J. Bouma and S. Gobert, Ontogenic variation and effect of collection procedure on leaf bio-mechanical properties of Mediterranean seagrass Posidonia oceanica (L.) Delile, *Marine Ecology* **37**(4), 750–759 (2016).

20. M. Paul and A.-J. Evertsen, Water temperature and salinity drive response to hydrodynamic forcing of Laminaria digitata. In *Proceedings of the 56th ECSA Conference,* Bremen, Germany (2016).

21. S. Vogel, Life in moving fluids: The physical biology of flow. Princeton University Press, Princeton, N.J (1994).

22. M.J.A. Christianen, J. van Belzen, P.M.J. Herman, M.M. van Katwijk, Lamers, Leon P. M., van Leent, Peter J. M. and T.J. Bouma, Low-canopy seagrass beds still provide important coastal protection services, *PLoS ONE* **8**(5), e62413 (2013).

23. B. Marin-Diaz, T.J. Bouma and E. Infantes, Role of eelgrass on bed-load transport and sediment resuspension under oscillatory flow, *Limnology and Oceanography* **65**(2), 426–436 (2020).

24. T. Spencer, I. Möller, F. Rupprecht, T.J. Bouma, B.K. van Wesenbeeck, M. Kudella, M. Paul, K. Jensen, G. Wolters, M. Miranda-Lange and S. Schimmels, Salt marsh surface survives true-to-scale simulated storm surges, *Earth Surface Processes and Landforms* **41**(4), 543–552 (2016).

25. M. Paul and N.B. Kerpen, Erosion protection by winter state of salt marsh vege-tation, *Journal of Ecohydraulics*, 1–10 (2021).

26. G. Mariotti and S. Fagherazzi, A numerical model for the coupled long-term evolution of salt marshes and tidal flats, *Journal of Geophysical Research* **115**(F1), 77 (2010).

27. Y. Chen, M.B. Collins and C.E.L. Thompson, Creek enlargement in a low-energy degrading saltmarsh in southern England, *Earth Surface Processes and Landforms* **36**(6), 767–778 (2011).

28. E. Mcleod, G.L. Chmura, S. Bouillon, R. Salm, M. Björk, C.M. Duarte, C.E. Lovelock, W.H. Schlesinger and B.R. Silliman, A blueprint for blue carbon: Toward an improved understanding of the role of vegetated coastal habitats in sequestering CO_2, *Frontiers in Ecology and the Environment* **9**(10), 552–560 (2011).

29. M.F. Adame, S. Cherian, R. Reef and B. Stewart-Koster, Mangrove root biomass and the uncertainty of belowground carbon estimations, *Forest Ecology and Management* **403**, 52–60 (2017).

30. H. Brooks, I. Möller, S. Carr, C. Chirol, E. Christie, B. Evans, K.L. Spencer, T. Spencer and K. Royse, Resistance of salt marsh substrates to near-instantaneous hydrodynamic forcing, *Earth Surface Processes and Landforms* (2020).

31. S.D. Addo-Danso, C.E. Prescott and A.R. Smith, Methods for estimating root biomass and production in forest and woodland ecosystem carbon studies: A review, *Forest Ecology and Management* **359**, 332–351 (2016).

32. C. Chirol, S.J. Carr, K.L. Spencer and I. Moeller, Pore, live root and necromass quantification in complex heterogeneous wetland soils using X-ray computed tomography, *Geoderma* **387**, 114898 (2021).

33. G. Gyssels, J. Poesen, E. Bochet and Y. Li, Impact of plant roots on the resistance of soils to erosion by water: A review, *Progress in Physical Geography* **29**(2), 189–217 (2005).

34. Y. Chen, C.E.L. Thompson, M.B. Collins, Y. Chen, C. Thompson and M.B. Collins, Saltmarsh creek bank stability: Biostabilisation and consolidation with depth, *Continental Shelf Research* **35**, 64–74 (2012).

35. Y. Chen, C. Thompson and M. Collins, Controls on creek margin stability by the root systems of saltmarsh vegetation, Beaulieu Estuary, Southern England, *Anthropocene Coasts* **2**(1), 21–38 (2019).

36. R.A. Feagin, J.L. Irish, I. Möller, A.M. Williams, R.J. Colón-Rivera and M.E. Mousavi, Short communication: Engineering properties of wetland plants with application to wave attenuation, *Coastal Engineering* **58**(3), 251–255 (2011).

37. T.I. Marjoribanks and M. Paul, Modelling flow-induced reconfiguration of variable rigidity aquatic vegetation, *Journal of Hydraulic Research*, 1–16 (2021).

38. M. Paul and C.B. de los Santos, Variation in flexural, morphological, and biochemical leaf properties of eelgrass (Zostera marina) along the European Atlantic climate regions, *Marine Biology* **166**(10), 2187 (2019).

39. C.M. Duarte, Temporal biomass variability and production/biomass relationship of seagrass communities, *Marine Ecology Progress Series* **51**, 269–276 (1989).

40. A. Silinski, K. Schoutens, S. Puijalon, J. Schoelynck, D. Luyckx, P. Troch, P. Meire and S. Temmerman, Coping with waves: Plasticity in tidal marsh plants as self-adapting coastal ecosystem engineers, *Limnology and Oceanography* **63**(2), 799–815 (2018).

41. J. Carus, M. Paul and B. Schröder, Vegetation as self-adaptive coastal protection: Reduction of current velocity and morphologic plasticity of a brackish marsh pioneer, *Ecology and Evolution* **6**(6), 1579–1589 (2016).

42. D.A. Friess, K.W. Krauss, E.M. Horstman, T. Balke, T.J. Bouma, D. Galli and E.L. Webb, Are all intertidal wetlands naturally created equal? Bottlenecks, thresholds and knowledge gaps to mangrove and saltmarsh ecosystems, *Biological reviews of the Cambridge Philosophical Society* **87**(2), 346–366 (2012).

43. S.A. Kelly, T.M. Panhuis and A.M. Stoehr, Phenotypic plasticity: molecular mechanisms and adaptive significance, *Comprehensive Physiology* **2**(2), 1417–1439(2012).

44. K.A. Moore and F.T. Short, Zostera: Biology, ecology, and management: Chapter 16. In A.W.D. Larkum, R.J. Orth, C.M. Duarte (Eds.), *Seagrasses: Biology, Ecology and Conservation*, pp. 361–386, Springer (2006).

45. M.M. van Katwijk and D.C.R. Hermus, Effects of water dynamics on Zostera marina: Transplantation experiments in the intertidal Dutch Wadden Sea, *Marine Ecology Progress Series* **208**, 107–118 (2000).

46. Y.A. La Nafie, C.B. de los Santos, F.G. Brun, M.M. van Katwijk and T.J. Bouma, Waves and high nutrient loads jointly decrease survival and separately affect morphological and biomechanical properties in the seagrass Zostera noltii, *Limnology and Oceanography* **57**(6), 1664–1672 (2012).

47. L.M. Soissons, M.M. van Katwijk, G. Peralta, F.G. Brun, P.G. Cardoso, T.F. Grilo, B. Ondiviela, M. Recio, M. Valle, J.M. Garmendia, F. Ganthy, I. Auby, L. Rigouin, L. Godet, J. Fournier, N. Desroy, L. Barillé, P. Kadel, R. Asmus, P.M.J. Herman and T.J. Bouma, Seasonal and latitudinal variation in seagrass mechanical traits across Europe: The influence of local nutrient status and morphometric plasticity, *Limnology and Oceanography* **35**, 269 (2017).

48. C.B. de los Santos, Y. Onoda, J.J. Vergara, J.L. Pérez-Lloréns, T.J. Bouma, Y.A. La Nafie, M.L. Cambridge, F.G. Brun, T.J. Bouma, Y.A. La Nafie, M.L. Cambridge and F.G. Brun, A comprehensive analysis of mechanical and morphological traits in temperate and tropical seagrass species, *Marine Ecology Progress Series* **551**, 81–94 (2016).

49. S. Puijalon, T.J. Bouma, C.J. Douady, J. van Groenendael, N.P.R. Anten, E. Martel and G. Bornette, Plant resistance to mechanical stress: evidence of an avoidance-tolerance trade-off, *The New Phytologist* **191**(4), 1141–1149 (2011).

50. D. Schulze, F. Rupprecht, S. Nolte and K. Jensen, Seasonal and spatial within-marsh differences of biophysical plant properties: Implications for wave attenuation capacity of salt marshes, *Aquatic Sciences* **81**(4), 82 (2019).

51. E. Carrington, Drag and dislodgment of an intertidal macroalga - Consequences of morphological variation in Mastocarpus-Papillatus Kutzing, *Journal of Experimental Marine Biology and Ecology* **139**(3), 185–200 (1990).

52. B. Gaylord, C.A. Blanchette and M.W. Denny, Mechanical consequences of size on wave-swept algae, *Ecological Monographs* **64**(3), 287–313 (1994).

53. C.A. Blanchette, Size and survival of intertidal plants in response to wave action: A case study with Fucus gardneri, *Ecology* **78**(5), 1563–1578 (1997).

54. F. Rupprecht, I. Möller, M. Paul, M. Kudella, T. Spencer, B.K. van Wesenbeeck, G. Wolters, K. Jensen, T.J. Bouma, M. Miranda-Lange and S. Schimmels, Vegetation-wave interactions in salt marshes under storm surge conditions, *Ecological Engineering* **100**, 301–315 (2017).

55. M. Paul and L.G. Gillis, Let it flow: How does an underlying current affect wave propagation over a natural seagrass meadow?, *Marine Ecology Progress Series* **523**, 57–70 (2015).

56. J.D. Ackerman and A. Okubo, Reduced mixing in a marine macrophyte canopy, *Functional Ecology* **7**, 305–309 (1993).

57. M. Taphorn, R. Villanueva, M. Paul, J. Visscher and T. Schlurmann, Flow field and wake structure characteristics imposed by single seagrass blade surrogates, *Journal of Ecohydraulics*, 1–13 (2021).

58. H.M. Nepf, Flow and transport in regions with aquatic vegetation, *Annual Review of Fluid Mechanics* **44**(1), 123–142 (2012).

59. R. Villanueva, M. Thom, J. Visscher, M. Paul and T. Schlurmann, Wake length of an artificial seagrass meadow: A study of shelter and its feasibility for restoration, *Journal of Ecohydraulics*, 1–15 (2021).

60. M. Paul, T.J. Bouma and C.L. Amos, Wave attenuation by submerged vegetation: Combining the effect of organism traits and tidal current, *Marine Ecology Progress Series* **444**, 31–41 (2012).

61. M. Paul, F. Rupprecht, I. Möller, T.J. Bouma, T. Spencer, M. Kudella, G. Wolters, B.K. van Wesenbeeck, K. Jensen, M. Miranda-Lange and S. Schimmels, Plant stiffness and biomass as drivers for drag forces under extreme wave loading: A flume study on mimics, *Coastal Engineering* **117**, 70–78 (2016).

62. I. Möller, M. Kudella, F. Rupprecht, T. Spencer, M. Paul, B.K. van Wesenbeeck, G. Wolters, K. Jensen, T.J. Bouma, M. Miranda-Lange and S. Schimmels, Wave attenuation over coastal salt marshes under storm surge conditions, *Nature Geoscience* **7**(10), 727–731 (2014).

63. K. Schoutens, S. Reents, S. Nolte, B. Evans, M. Paul, M. Kudella, T.J. Bouma, I. Möller, S. Temmerman, K. Schoutens, S. Reents and T. Bouma, Survival of the thickest? – Impacts of extreme wave-forcing on marsh seedlings is mediated by species morphology // Survival of the thickest? Impacts of extreme wave-forcing on marsh seedlings are mediated by species morphology, *Limnology and Oceanography* (2021).

64. J. Carus, M. Heuner, M. Paul and B. Schröder, Which factors and processes drive the spatio-temporal dynamics of brackish marshes? – Insights from development and parameterisation of a mechanistic vegetation model, *Ecological Modelling* **363**, 122–136 (2017).

Chapter 3

Effect of Coastal Vegetation in Attenuating Extreme Waves – A Review

S. Harish[1,2], V. Sriram[2,*], B. Jochems[1] and Holger Schüttrumpf[1]

[1]*Institute of Hydraulic Engineering and Water Resources Management, RWTH Aachen University, Aachen, North Rhine-Westphalia, Germany*
[2]*Department of Ocean Engineering, Indian Institute of Technology Madras, Adayar, Chennai, Tamil Nadu, India*
[*]vsriram@iitm.ac.in

The past tsunami events, despite inducing extensive damage to the coastal community, cognized coastal vegetation as a buffer to the inundating tsunami. This, as well as the endangerment of coastal forest as ecosystems, necessitates a deeper understanding of coastal vegetation's potential for tsunami damage mitigation. The literature highlighted the energy reduction of up to 60% due to the presence of vegetation during tsunami inundation conditions. Thus, several studies in the past and recent times investigated the effect of vegetation on tsunami energy reduction. This chapter extensively reviews in detail the vegetation parameter and flow parameter influence on energy reduction with a review furthering on the hybrid defence system (combination of vegetation and hard structures). The chapter presents a clear idea for building a coastal forest, thereby helping identify the best combination of the parameters for availing the maximum advantage of the vegetation with the available land space.

Keywords: Coastal forest, vegetation, tsunami mitigation, hybrid defence system.

[*]Corresponding author.

1. Introduction

The increase in the frequency of occurrence of extreme coastal events necessitates improved coastal resilience. The coastal resilience to one such extreme event, called a tsunami, is the topic of interest in this chapter. In past tsunami events, many of the hard coastal protection structures built along the coast failed and were found to offer insufficient protection to the coast[1,2]. Consequently, many different studies have been conducted, aiming to improve the capability of protecting coastal regions in the event of tsunami inundation[3–7]. While analyzing the failure modes of dikes after the 2011 Tohoku Oki tsunami, it became apparent that natural obstacles behind the embankment, such as trees, can play a major role in damage mitigation[8,9]. For one thing, every obstacle in the way of the water flow imposes drag and thus reduces its hydrodynamic energy. But the increased resistance also results in a greater inundation height right behind the embankment (i.e., in front of the vegetation), which changes the motion pattern of the water and thus helps prevent damage to the embankment due to scouring[8]. Apart from the effect of the vegetation on the free surface bore, a coastal forest offers other benefits as well. An appropriate combination of smaller (i.e. *P. odoratissimus*/Screw-pine) and larger trees (i.e. *C. equisetifolia*/Coastal She-oak) can trap debris, serve as an escape route from the flood, and provide a soft landing for people already washed away[10].

While the topic of vegetation in tsunami defence systems – either solo as a single defence system (SDS) or in combination with an embankment as a hybrid defence system (HDS) – has been a subject of research for more than two decades, recent tsunami disasters especially have been an inducement for many studies. A decent understanding of possible ways a forest can contribute to damage mitigation has been developed[9,10], as well as the interactions between an embankment and vegetation. Similarly, the influence of tree size and spacing on their ability to withstand free surface bore and dissipate hydrodynamic energy, even which species are especially suitable for the task, is now well understood[10]. It was concluded that neither an embankment nor coastal vegetation could offer sufficient protection alone[11] (see Fig. 1). While a forest's ability to withstand the force of a giant tsunami is limited without the physical protection of an embankment[12], a vegetation patch can help to protect the dike from scouring when overflowed[13]. Culturing coastal vegetation offers the additional benefit of being less resource-intensive compared to constructing a traditional embankment[14,4,15].

Fig. 1. Damage to overflowed embankments after the 2011 Tōhoku Tsunami[8].

To maximize the protection of human lives and settlements, it is thus desirable to implement vegetation along with the traditional tsunami defence systems (i.e., hard structures), as well as deliberately use and foster existing coastal forests for their potential to mitigate tsunami damage. A set of vegetation parameters have been proven to significantly impact energy reduction and was thus instrumental in the design of coastal vegetation. The current knowledge about vegetation parameters and their influence on tsunami energy reduction is summarized in the following sections. Further, the present chapter briefly discusses the HDS system and its significance. The expected scale effects in the vegetation experiments are further discussed, and finally, the conclusions of this review chapter are projected.

2. Vegetation Parameters

It is crucial to identify the characterizing parameters to model a coastal forest. Concerning the vegetation design, a set of geometric parameters have been established by previous research, and, further, the general arrangement of trees. These parameters include the vegetation thickness, the spacing and the diameter of trees, and the width of the vegetated area.

2.1. *Thickness and Arrangement of Coastal Vegetation*

Past studies considered two types of tree arrangement: a staggered or a tandem pattern. Previous research has either found no significant difference[8] between the efficacy of these patterns or determined the staggered arrangement to be more effective[16,17].

Different approaches can be found in the literature to express the forest thickness. Ref. 18 introduced the summed tree diameter as a distinct parameter to describe the thickness of the vegetation.

"d_n is the 'summed tree diameter', which is defined by a product of the breast diameter and the number of trees in a rectangle with a frontage of unit length along the shoreline and a depth equal to the width of the forest.[18]"

The summed tree diameter for the staggered arrangement can be calculated as

$$d_n = \frac{2}{\sqrt{3}D^2} W_v d \tag{1}$$

where d_n [No.cm] is the summed tree diameter, d [cm] is the tree diameter in cm and D [m] is the centre-to-centre distance between two trunks perpendicular to the direction of the flow. Herein, the first term ($2/\sqrt{3}D^2$) represents the cylinder density in the staggered arrangement. d_n is kept as cm to adjust it to Ref. 18 representation. Since d_n is cumulating all trunks in the way of the bore, it is proportional to the width W_v [m] of the forest. Therefore, a longer forest of the same density will have a higher thickness d_n. The parameter is thus helpful to describe the overall resistance vegetation that can impose on the water flow. However, due to its dependency on both the width of the forest and its spacing, the same value d_n can be achieved with different spacings by varying the width. Consequently, the summed-up tree diameter cannot directly convey how much space is in between trees. Field surveys have been conducted to determine the vegetation thickness of coastal forests in nature. They were found to exhibit a thickness d_n of 130 to 400 [No.cm][8,19]. Modelled vegetations with a summed tree diameter in this range have been used in several previously conducted experiments[16,13,20–22].

Another approach commonly used is vegetation porosity. This value describes how much of an area is "free", meaning not occupied by trunks. The porosity of a forest can be calculated as

$$\mathrm{Pr} = 1 - \frac{n\,\pi\,\frac{d^2}{4}}{A} \qquad (2)$$

where Pr [%] is the porosity and n [-] is the number of trees with the diameter d [cm] in a vegetation patch of area A [m^2]. The porosity does not depend on the width of the forest, only on the spacing of the trees and their diameter. It can, therefore, convey a sense of sparseness or denseness in the vegetation. In previous research, the modelled forests mostly featured a porosity of 0.93 to 0.99 [8,16,13,23,21,22,20].

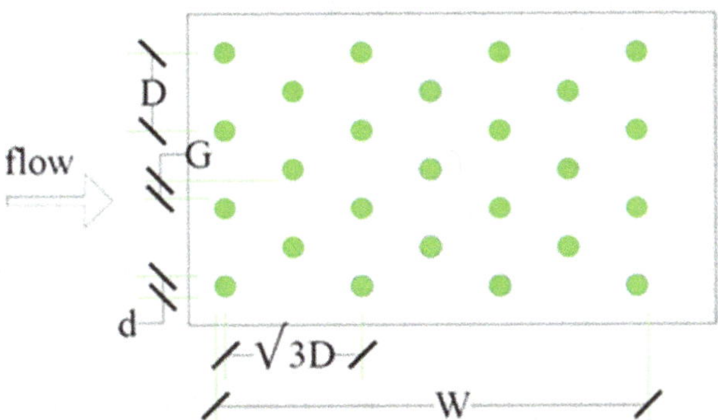

Fig. 2. Definition of vegetation parameters in plan view.

The spacing can also be described as the ratio of the edge-to-edge distance between two trees cross-directional to the flow G [m] and their diameter d [m] (Fig. 2). This value also does not indicate the total water resistance of the forest, only its density. Because this parameter is directly derived from the free space between trees and their diameter, it is most accessible and can be easily visualized. In previous studies, experiments with G/d values from 0.25 to 5.25 [-] have been conducted[24–26]. While low G/d values of approximately $0.25 - 1$ [-] are not applicable in a real forest, as the trees need enough light, the forests with values higher than 2 can be realized using different species of trees and offer a high potential for energy reduction[27,23].

2.2. *Spacing of Trees and Trunk Diameter*

After the 2004 Indian Ocean tsunami, a field survey assessed damage to vegetated shores. Parameters such as the species of trees, spacing, trunk diameter and suffered damage were catalogued, resulting in a deeper understanding of the trees' morphology and its influence on their ability to withstand a tsunami bore.

From a hydrodynamical point of view, it is desirable to have vegetation of great density, given that the increased drag will result in improved energy dissipation, but it is limited in practice. Depending on its size, a tree needs enough light to grow properly. In natural forests, a specific spacing depending on the trunk's diameter can therefore be observed, also varying for individual species[10]. Coastal vegetation of high density for maximum energy reduction thus necessitates small trees with accordingly thin trunks.

"[...] A larger tree requires a larger spacing (lower tree density) to grow. Therefore, the effect of [the diameter] d and [spacing] D cannot be discussed independently[10]."

However, if a trunk is too thin, it is often found unable to withstand the force of the oncoming water or impacting debris. If not sufficiently rooted, a tree can even be overturned entirely[28]. In addition, broken or uprooted trees not only fail to constitute the desired water resistance but also generate floating debris. This implies the necessity for greater trunk diameter trees. However, bigger trees need more space, resulting in reduced forest density. Consequently, a compromise between those two factors must be found. To account for that, previous research recommended trees with a diameter of 0.1 to 0.3 meters to withstand a level 1 tsunami [10,11], where level 1 tsunamis describe a tsunami event with a return period of 50–160 years having inundation depth lesser than 7–10 m. Figure 3 illustrates the relation between tree spacing and trunk diameter for various species typically found in the coastal zone.

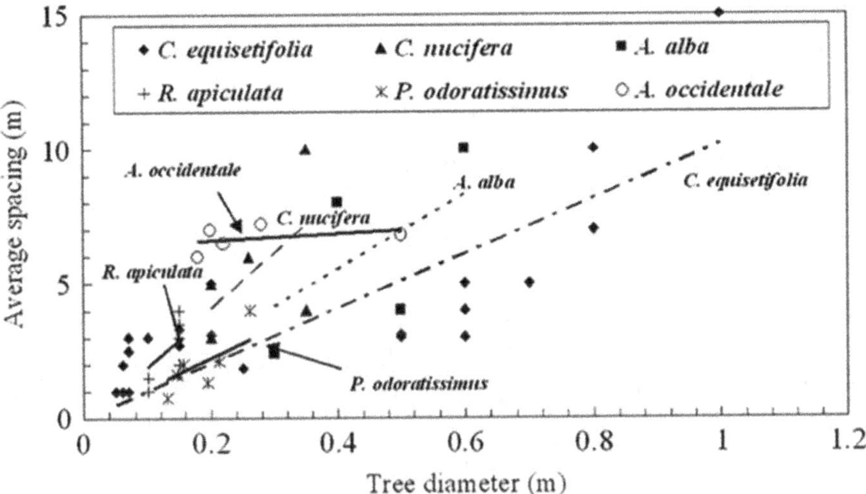

Fig. 3. Relation between spacing and trunk diameter[10].

2.3. *Forest Width*

Apart from the porosity, the width of the forest is another key factor of the resistance induced by the vegetation. Since the lowest possible porosity is limited for the above-mentioned reasons, it is interesting to further research the influence of the width of the coastal forest on energy reduction in case of a tsunami. Previous research on energy reduction was often carried out with a constant vegetational thickness d_n, varying the porosity and only, therefore, changing the width of the vegetation as well[21,26]. Ref. 8 carried out a study that did vary the width while keeping a constant porosity. However, it only examined the influence of the hydraulic jump type, not the energy reduction. A few studies[24,29–31] also conducted experiments by varying the vegetation width for a constant porosity to determine the energy reduction.

3. **Energy Dissipation due to Vegetation**

The goal of every tsunami defence system is to protect land and people in the direction of flooding. However, it is mostly impossible to completely contain the tsunami bore and prevent the inundation of cultivated land and settlements, especially for larger events. To evaluate the efficiency of a tsunami mitigation measure, it is necessary to introduce parameters describing the remaining aspect of the bore. The most commonly used parameter is the total energy of the bore, which is the sum of its potential and kinetic energy and can be calculated as

$$E = h + \alpha \frac{v^2}{2g} \tag{3}$$

where E [m] is the specific energy head of the flow, h [m] is the flow depth, v [m/s] is the velocity, and g [m/s^2] is the gravitational acceleration. The coefficient α accounts for variations in velocity along the flow. Under the controlled experimental conditions with a constant cross-section and discharge, the velocity coefficient (α) can be assumed as unity. This was stated by Ref. 32 and adopted by the previous studies[13,21]. By calculating the specific energy at two points – one upstream and one downstream of the vegetation – the reduction of energy can be expressed as

$$\Delta E \; [\%] = \frac{E_1 - E_2}{E_1} \times 100\% \tag{4}$$

where ΔE [%] is the relative energy loss and E_1 [m] and E_2 [m] are the energy head on the upstream and downstream side of the vegetation, respectively.

More research has been conducted in the past to analyze and quantify the process of energy reduction caused by vegetation. However, the focus has mostly been on specific energy change rather than momentum change. Two key results have been established: First of all, energy can either dissipate inside the vegetated area due to the drag imposed by the trunks or behind the vegetation by creating a well-controlled hydraulic jump[24]. Secondly, the amount of reduced energy depends on the parameters defining coastal vegetation and water flow, mainly the thickness of vegetation (or summed-up diameter) and the Froude number (Fr). Thicker vegetation leads to greater energy dissipation. The energy reduction inside the vegetation generally accounts for the larger part, but a well-regulated hydraulic jump can make a significant contribution [33,20]. Therefore, such a design is especially useful in places with limited space, where a forest of the necessary thickness cannot be constructed[24].

According to existing literature, a vegetation patch can reduce the hydrodynamic energy of the water flow by 20 to 60%. This depends on the thickness of vegetation, the density of the vegetation and the Fr of the flow, as already stated. In the following, the influence of different parameters is further described.

3.1. *Influence of the density (G/d)*

Experiments with constant Fr and vegetation thickness but varying *G/d* were conducted to isolate the influence of the density of the vegetation. Here, *G* corresponds to spacing. It should be noted that the varied parameter was the spacing *G*, whereas the tree diameter *d* was not changed. As Ref. 24 observed, the drag force on the trees increased with increasing density. In this case, however, the increased drag clearly corresponded to an increased energy loss. The experiments featured vegetation models of three different thicknesses (d_n = 180, 380, 580), which were each tested in three different density configurations (*G/d* = 0.25, 1.09, 2.13 [-]). Those were also tested for a range of Fr (0.57 [-] – 0.73 [-]). By way of example, the results for Fr = 0.68 are represented in Fig. 4. A total difference of 5 to 9% in relative energy loss was measured

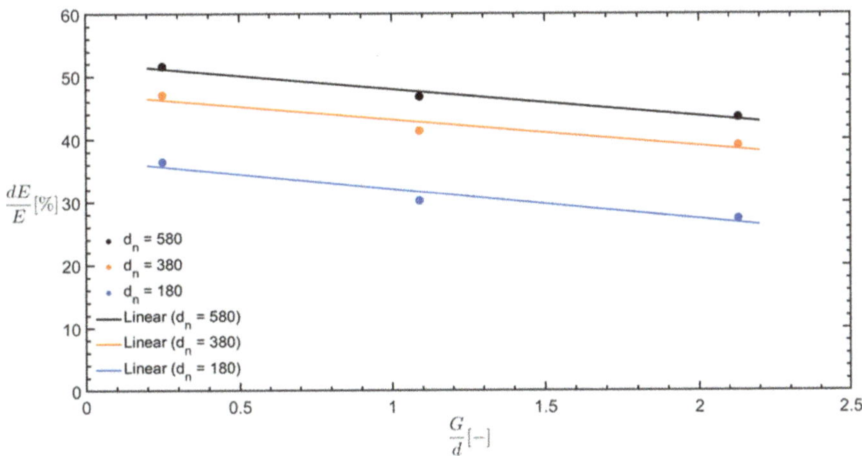

Fig. 4. Influence of density *G/d* on relative energy loss for Fr = 0.68 (own representation based on Ref. 24).

between dense (G/d = 0.25 [-]) and sparse (G/d = 2.13 [-]) vegetation. Ref. 5 observed that the velocity behind the vegetation was reduced by about 17% when the density decreased from G/d = 4.0 [-] to G/d = 0.0 [-] for the constant vegetation thickness. This indicates that the forest should be as dense as possible to achieve maximum energy reduction, which is, nevertheless, influenced by the minimum spacing required for the trees to grow, as discussed above.

3.2. *Influence of the vegetation thickness*

In experiments with varying thicknesses but constant Froude number and density, it became apparent that increasing d_n results in increasing energy reduction[24]. The experiments conducted by Ref. 24 featured vegetation models of three different densities (G/d = 0.25, 1.09, 2.13), which were each tested in three different thickness configurations (d_n = 180, 380, 580). Those were also tested for a range of Froude numbers. Here, the sample results for Fr = 0.68 are represented in Fig. 5.

While increasing the vegetation thickness from d_n = 180 to 380 No.cm, an increase in the energy reduction of 9 to 12% was measured. Further, increasing the thickness to d_n = 580 No.cm resulted in an additional

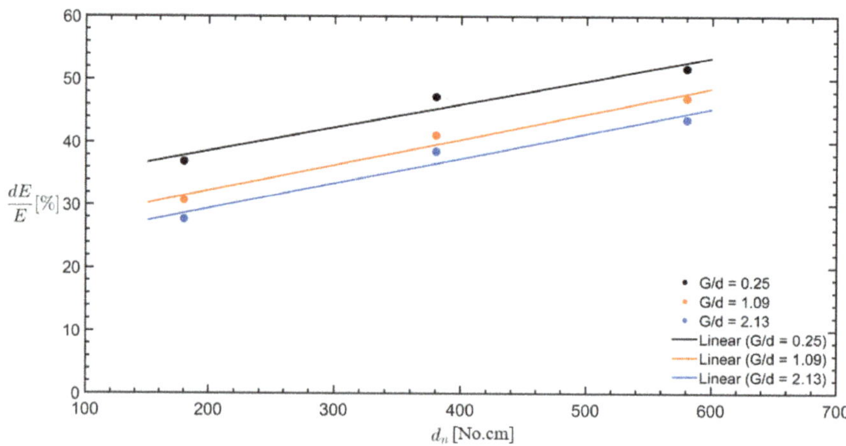

Fig. 5. Influence of thickness on relative energy loss for Fr = 0.68 (Own representation from Ref. 24) (d_n = 180 to 580 No.cm).

increase in the energy reduction of 5 to 6%. This gives an overall increase in the energy reduction of 15 to 17% between $d_n = 180$ No.cm and $d_n = 580$ No.cm. The vegetation thickness can therefore be considered the most significant parameter since it has the greatest influence on a coastal forest's capability to achieve total relative energy dissipation. Nevertheless, a larger vegetation thickness for a particular density requires a more landward space.

3.3. *Influence of the Froude number*

To isolate the effect of the flow's properties expressed through the Froude number, it is necessary to keep other parameters constant. The following studies conducted experiments with constant density and thickness while varying Fr. Ref. 21 tested a configuration of sparse and intermediate vegetation under varying flow conditions using merely a constant vegetation thickness ($d_n = 190$ No.cm). For sparse and intermediate vegetation of $G/d = 2.13$ and $G/d = 1.09$ in a flow with initial Froude number Fr = 0.40, the maximum energy reduction was 28% and 36%, respectively. The Froude number was then increased to 0.65, which caused a decrease in relative energy loss (23% and 27% for sparse and intermediate vegetation, respectively). The influence of the Froude number on the relative energy loss in sparse and intermediate vegetation, according to this study, is presented in Fig. 6.

The experiments featured a vegetation model in two different configurations, each tested in a flow with a Froude number ranging from 0.40 to 0.65. Since it was not independently tested against a broader range of both density and thickness, the above-mentioned literature only schematically represents the decline of energy loss due to an increasing Froude number for a particular density and thickness.

Ref. 24 observed a slightly increasing drag coefficient on the trees with an increasing Fr. Yet this did not seem to have a significant effect since the relative energy loss remained almost constant. It should be noted, however, that the initial Froude number only varied in the small range of 0.58 to 0.73 [-]. Similarly, the energy loss was found to increase with the increase in the Fr by Ref. 20. The energy loss increased by about 8% between Fr between 0.65 and 0.72. While testing in high Fr (1.08-1.56)[13],

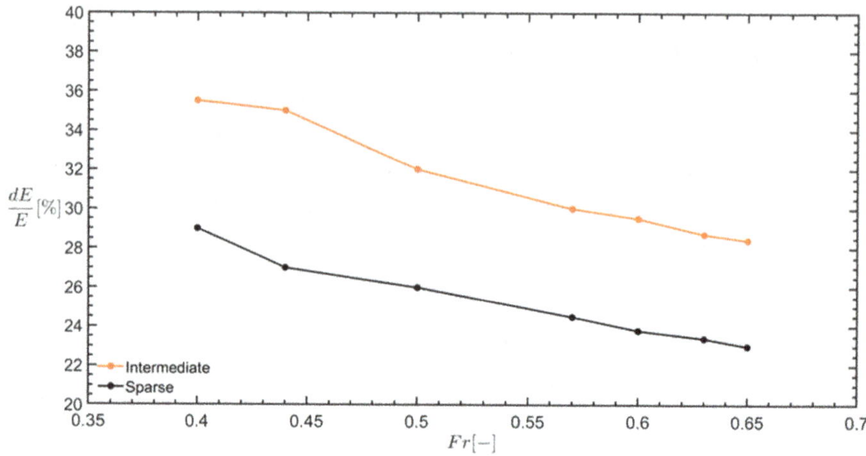

Fig. 6. Influence of Froude number on relative energy loss (own representation based on Ref. 21).

the energy loss increased up to the Fr of 1.44 and decreased suddenly. Hence, randomness is involved in the energy reduction variation with Fr, which must be evaluated in future studies.

Thus, the studies observed a clear dependence of the Fr on the energy dissipation in the vegetation system. Until now, the Fr tested with the vegetation was mainly in the near-critical and sub-critical regime[24,21,20,34] with a few recent studies in the super-critical regime [13,29,30,26,35]. Nevertheless, a unified equation relating the effect of Fr with the vegetation parameters in the sub-critical to the super-critical regime is not reported. This should be investigated in the future. Furthermore, for implementation, a thorough analysis of Fr expected at a specific location is essential to design the vegetation belt or to evaluate the energy reduction.

4. Energy Reduction in a Hybrid Defence System

The hybrid defence system encompasses a combination of hard and soft measures (Fig. 7). Although the discussion above portrays a clear sign of having coastal vegetation in the energy reduction, part of the forests were found destroyed after the 2004 IOT and 2011 TOT[10,36,11,37]. Despite being broken, the vegetation still reduced the tsunami energy[37], although the emergent and unbroken vegetation yielded the most advantage[11]. Ref. 21 and 13 noticed that adding an embankment reduced the energy by about 30-40% compared to the only vegetation case. The field observation after the 2011 TOT also confirms that the combination of the embankment and the vegetation in Sendai city resulted in lesser damage in the area behind the vegetation compared to the single defence alone[11,38]. Ref. 39 also proposed a defence system combining the coastal forest with a moat (deep, wide ditch or dug pools).

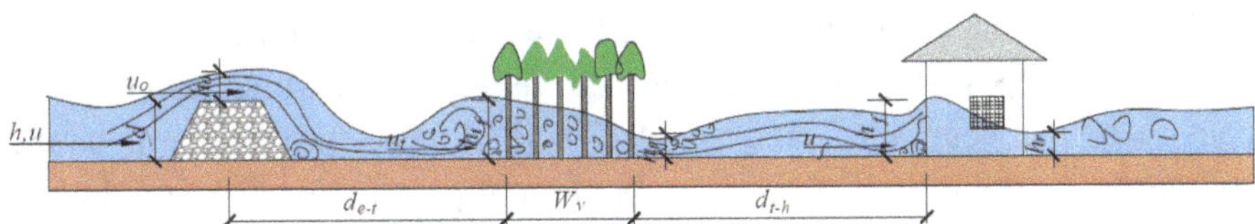

Fig. 7. Schematics of the hybrid coastal defence system.

Many studies have emerged to understand the significance of a hybrid defence system by combining the coastal forest and a moat[40], the sea embankment and the coastal forest[11,16,38], and the embankment, moat and forest[29,30,28]. Theoretically, increasing the number of resistance factors would decrease tsunami energy. Hence, the combination of the embankment, moat, and forest will perform well compared to the other systems. Depending upon the location of the existing forest or the existing sea embankments, the addition of the other defence structure order would vary[11]. Ref. 29 arrived at the best arrangement of the three coastal defences for maximum advantage through physical experiments. From the study with the dam-break bore, it was concluded that the hybrid system should be arranged in the order of vegetation (V), moat (M) and embankment (W) in the case of seaside vegetation or W+M+V in the case of landward vegetation. Nevertheless, the maximum reduction in the overtopping volume was observed in the former case since the energy of the fluid is greatly reduced due to the vegetation, hence the Fr, which reduced the overtopping[41]. However, the fluid force reduction was found to be maximum in the latter case. Further, the moat (M) alone did not significantly contribute to the fluid force reduction (as also observed by Ref. 40 from numerical simulations), where the fluid force was estimated as (hu^2), with

h and u being the measured bore height and velocity. Followed by Ref. 29, Ref. 30 conducted a similar set of experiments with different vegetation combinations in the hybrid defence system (emergent, submerged and in mixed configuration). Even in the submerged condition, the order of arrangement of each defence system for the maximum advantage remained the same, as discussed above. However, in the submergence condition, the flow reduction capacity (i.e., the inundating volume and the fluid force reduction) of the vegetation is reduced compared to the emergent condition[30,11].

Overall, the energy dissipation or the fluid force reduction of any hybrid defence system depends on the vegetation parameters (i.e., the density of the vegetation, the vegetation thickness, the emergent or the submergence condition) and the embankment parameters (i.e., the height of the embankment, the shape of the embankment, the distance between the embankment and the landward obstruction, Ref. 42). Furthermore, the distance between the embankment and vegetation was also found to influence the hydraulic jump characteristics[13,34] and hence the energy dissipation[24,27].

5. Scaling

The vegetation in the coastal zone is often larger (e.g., *Cocos nucifera, Casuarina equisetifolia, Pandanus odoratissimus, Rhizophora species*); hence the hydrodynamic analysis under laboratory conditions is performed in the scaled-down model. Few studies modelled the mangrove forest (common along many coastal cities of India, Indonesia and Srilanka) in which the mangroves are modelled using plastic or wire roots with a similar porosity as the root system[43–45]. With an adopted model scale for the vegetation, the actual behaviour of inundating bore must also resemble that of the prototype. Since all physical processes must be effectively recreated during the laboratory study, it is essential to evaluate the various forces acting on the test model and ensure that all physical processes are analogous between the prototype and the model[46].

Different model scaling laws could be used depending on the type of study being conducted to ensure similarity between the prototype and the laboratory models. In the problems concerning fluid flows, the forces acting on the system may be any of the below or a combination of different forces.

i) Inertial forces
ii) Frictional or viscous forces
iii) Gravity forces
iv) Pressure forces
v) Elastic force
vi) Surface tension forces

The main restoring forces in the tsunami vegetation interaction studies are the gravity force and the elasticity of the vegetation. The skin-friction drag force can be considered negligible. Most of the laboratory modelling of tsunami vegetation interaction studies is usually carried out by using Froude's scaling law, as can be seen in the work of Refs. 29,21,24,27,11.

5.1. *Froude scaling law*

The Froude scaling relates the inertial forces to the gravitational forces as follows.

$$\sqrt{\frac{inertial\ force}{gravity\ force}} = \sqrt{\frac{\rho L^2 u^2}{\rho L^3 g}} = \frac{u}{\sqrt{gL}} \tag{5}$$

where 'u' and 'g' are the flow velocity and the gravitational constant, respectively, and 'L' is the characteristics length, which can be taken as the flow depth 'h' in the open-channel flow conditions.

The Froude scaling law should be applied only when the predominant reaction force on the system is due to gravity and inertia, which governs the fluid flow. Applying this law in the physical model study requires that the Froude number in the prototype must be equal to the Froude number in the model.

$$\left(\frac{u}{\sqrt{gL}}\right)_p = \left(\frac{u}{\sqrt{gL}}\right)_m \tag{6}$$

where the subscripts 'p' and 'm' denote the corresponding Froude number at the prototype and the model scale.

     S. Harish, V. Sriram, B. Jochems and H. Schüttrumpf

As given in Eqn. 7, the scaling parameter 'Δ' is defined as the ratio of the prototype characteristic length to the length in the model scale. In this way, all the similitude parameters of the Froude scaling law can be defined.

$$\Delta = \frac{L_p}{L_m} \tag{7}$$

Based on Froude's scaling law, the main parameters used and its corresponding scale factor are given in Table 1.

Table 1. Froude Scaling conversion factor

Variable	Unit	Scale factor
Length	[m]	Δ
diameter	[m]	Δ
Flow depth	[m]	Δ
Velocity	[m/s]	$\sqrt{\Delta}$
Time	[s]	$\sqrt{\Delta}$
Time period	[s]	$\sqrt{\Delta}$
Force	[N]	Δ^3
Mass	[kg]	Δ^3
Flexural rigidity	[Nm^2]	Δ^5
Pressure	[Pa=N/m^2]	Δ
Moment	[Nm]	Δ^4

5.1.1. *Scale effects*

Even though physical modelling implements the Froude scaling law, the secondary forces which are neglected during model scaling may not be scaled correctly. Since the Froude scaling law includes the predominant reaction force as gravity force, the effects of the exclusion of viscosity, elasticity and surface tension can cause inaccuracy in determining the hydrodynamic behaviour of the system[47]. Nevertheless, it is impossible to replicate both gravity-related and viscous processes at the same time. These issues cause the non-similarity between the prototype and the model, and its effects on predicting prototype response are called scale effects.

At first, during bore propagation in the laboratory, the tip of the bore is associated with a lot of turbulent aeration mix due to shear flow between the quiescent water and the high-velocity flow[41]. Secondly, during interaction with the vegetation, the vegetation also induces turbulence in the flow field[34,48], with much aeration in the vegetation submerged condition[30]. Thirdly, the flow, after passing the vegetation, could result in a hydraulic jump which also induces aeration[24,13]. At last, most experiments were carried out with fresh water than with seawater, expecting higher aeration for seawater than for the freshwater used in the laboratory experiments[49]. To all this problem, modelling aeration in the flow field is difficult without considering the surface tension and air bubble dynamics[50,51]. For such an aerated bore condition, Cauchy's number (inertial force/elastic force) could be used. Nevertheless, Ref. 49 concluded that the Froude scaling law could still be used reasonably for the force or pressure scaling since Cauchy's law could result in underestimation of the impact pressure, thus suggesting the usage of the Froude law even for such aerated flows. A minor difference in energy dissipation is always expected between the model and the prototype, and this effect would decrease when the experiments are conducted on a large scale[49].

To minimize the effect of viscosity, the Reynolds number (Re) must be sufficiently large[52,50]. Hence, for example, if an experiment is carried out at 1:100 in the Fr scale, the Reynolds number is 1000 times smaller (assuming constant density and viscosity of the fluid between the model and prototype scale) in the prototype compared to the real scale. Especially for hydraulic jumps, Ref. 53 suggested a $Re > 10^5$ for $Fr < 10$ to minimize scale effects. However, many studies were still conducted in the low Re range. Refs. 52 and 54 suggested a minimum wave height of 2 cm and Weber number (We) > 11 to avoid the scaling effect due to surface tension. Thus, in any experiments, a minimum flow depth of 2 cm and above has to be maintained to avoid surface tension effects. Table 2 summarizes the limiting criteria for flow parameters to avoid scale effects.

Table 2. Limiting criteria to avoid significant scale effects in wave-vegetation experiments

Dimensionless number	Value	Comments
Reynold's number	$>10^5$	For Fr < 10
Weber number	> 11	Wave height > 2 cm wave period > 0.35 s

5.2. *Modelling of vegetation trunk*

The vegetation should be modelled for elasticity to replicate the real flow-vegetation interaction problem[46].

As per Froude's modelling for the force,

$$\frac{F_p}{F_m} = \left(\frac{\gamma_{s,w}}{\gamma_{f,w}}\right)\left(\frac{L_p}{L_m}\right)^3 \qquad (8)$$

Where γ is the specific weight. Subscripts 's,w' and 'f,w' represents the seawater and the fluid tested in the laboratory. In terms of the scale ratio for force (N_F),

$$N_F = \left(\frac{\gamma_{s,w}}{\gamma_{f,w}}\right)(\Delta)^3 \qquad (9)$$

The elastic property of the tree can be modelled as per Cauchy's modelling criteria[46], which is given as

$$\frac{F_{Ep}}{F_{Em}} = \frac{(EA)_p}{(EA)_m} \qquad (10)$$

Where F_E is the elastic force, E is young's modulus, and A is the cross-sectional area. In terms of the scale ratio for the elastic force for un-distorted models,

$$N_{F,E} = \left(\frac{E_p}{E_m}\right)(\Delta)^2 \qquad (11)$$

The elastic force scale must be the same as the hydrodynamic force scale. Equating the Fr criteria and Ca criteria and requiring that the tree is geometrically undistorted,

$$\left(\frac{E_p}{E_m}\right) = \left(\frac{\gamma_{s,w}}{\gamma_{f,w}}\right)\Delta \tag{12}$$

Which is the elasticity requirement for selecting model trees that correctly model elastic effects in the linear strain range resulting from the hydrodynamic loading in a Froude-scaled model. Similarly, the specific gravity of the model material chosen as per Fr law should be

$$\frac{S_p}{S_m} = \left(\frac{\gamma_{s,w}}{\gamma_{f,w}}\right) \tag{13}$$

In the case of hydrodynamic force, the force depends on the incoming fluid property. Hence, the γ_p can be considered the specific weight of the seawater and γ_m is the specific weight of the fluid examined in the laboratory. Assuming, $\gamma_p = \gamma_m$, the elastic property of the model and the specific gravity of the model is

$$E_m = \frac{E_p}{\Delta}; \ S_m = S_p \tag{14}$$

In reality, the density of the timber can be in the range of 10 to 20 GPa, and the specific gravity can range between 0.35 to 0.5. Thus, if a geometric scale of 1:100 is applied, the elastic model should have Young's modulus ranging between 0.1 to 0.2 GPa and the specific gravity between 0.35 and 0.5. In reality, finding the material possessing the above-required property is hard.

Since only the bending action of the model is of primary importance, alternatively, few studies[55,17,56,48] have modelled the rigidity (Δ^5) using Froude scaling (*EI*, where *I* is the moment of inertia in m^4) of the vegetation to model the bending action. This would result in a slightly different diameter of the vegetation trunk compared to the geometric similitude depending on the material chosen. Nevertheless, Refs. 48,57, 55,17,56 have applied it only to the base diameter of the vegetation to replicate the natural mode of vibration of the trees so that the geometric similitude is preserved.

5.2.1. *Model effects*

Nevertheless, many experimentally downscaled vegetation models are bounded to model effects. The tree is usually modelled smooth and not modelled for surface roughness. Further, the trunk of the tree is modelled rigidly. Thus, they are usually designed to simulate the flow obstruction without considering the response of the vegetation. The roughness could hardly influence the results in the quasi-steady flow phase, and hence its effect can be considered negligible. However, Refs. 17 and 55 identified that the vortex-induced vibration of the flexible vegetation could partially influence energy dissipation. Hence, experiments modelling the flexural rigidity of the vegetation could be of interest to future studies. In addition, modelling the complex root system is still challenging, and researchers proposed different vegetation models for modelling the complex root system[45,58,44]. Accurate field modelling is complicated, considering the randomness in the natural growth of the system. In addition, most experiments assumed that the flow interacts only with the trunk portion of the tree[24,29]. The influence of the branches and leaves of the trees is not modelled and hence represents an idealized condition.

6. Concluding Remarks and Outlook

Following the 2004 Indian Ocean tsunami, there has been a significant increase in the studies on vegetation as an effective source of tsunami energy reduction. This chapter extensively reviewed the effect of vegetation parameters in the single defence system (SDS) and the hybrid defence system (HDS) on tsunami energy reduction. The literature review identified a strong dependence on vegetation parameters (vegetation thickness, density and submerged or emerged conditions) and flow parameters (Froude number) on tsunami energy reduction. In the case of HDS, the embankment height, the location of the embankment and the distance between the vegetation are the influencing factors. The research contribution of Prof. Norio Tanaka (Saitama University) to the above-mentioned research is worth mentioning. The separate contribution from his group in this book chapter can be seen in.

Despite a significant number of earlier studies on vegetation, there are some research gaps which will be of interest to future studies.

- Most studies did not consider the influence of trees' branches in the experiments, especially in submerged conditions. The branches could significantly influence energy reduction since it contributes to the additional resistance from the trees.
- Studies on sparse vegetation ($G/d > 4$) should be increased since sparse vegetation allows people to flee through it easily during an extreme scenario compared to dense vegetation. Sparse vegetation would be of interest, especially in the populated coastal areas, without losing the aesthetic view of the beach.
- Studies shall focus on understanding the effect of flow Froude number by experimenting over a wider range of Fr in the range of previously occurred tsunamis; thereby, a unified equation for energy reduction shall be developed, combining vegetation parameters and the flow parameters.
- Most studies arrived at the fluid force reduction directly through the velocity (u) and bore depth (h) measurements (i.e., fluid force-hu^2). Studies shall confirm the fluid force reduction by the force measurements on a structure (say buildings). This is because the reflected bore from the structure could partially influence the flow characteristics inside the vegetation[35].
- The effect of seawall height in the HDS system should be focused further to arrive at an optimized vegetation design.
- Studies on HDS should further investigate different types of hard structures, including rubble mound seawalls, vertical seawalls, and recurve walls which are more common along different coastal cities. Furthermore, combining vegetation with artificial elements, such as *buffer blocks*, for tsunami energy dissipation could be a promising approach, considering its effectiveness in enhancing energy dissipation of wave overtopping in existing dikes[59].

- The vegetation experiments shall be conducted at a larger model scale compared to the previous studies to identify the potential influence of viscous forces and surface tension (scale effects) since most past experiments featured a Fr scale of 1:100.

Overall, the significance of vegetation is now well-established in tsunami energy reduction. With the available information through research experiments and numerical simulations, it is thus possible to evaluate the tsunami mitigation offered by the vegetation for the safe development of the infrastructure at the coast.

Acknowledgements

This work is partially supported by the Department of Science & Technology, India Grant No. DST/CCP/CoE/141/2018C under SPLICE – Climate Change Programme. This contribution also received funding from the German Research Foundation (Deutsche Forschungsgemeinschaft, DFG) for the project "Buffer blocks as wave energy dissipators (BB-WEnDis)" (SCHU 1054/21-1). This work is also contributed from the Center For Large scale Ocean Research (CFLOR), IITMadras.

References

1. N. Nandasena, Y. Sasaki and N. Tanaka, Modeling field observations of the 2011 Great East Japan tsunami: Efficacy of artificial and natural structures on tsunami mitigation, *Coastal Engineering.* **67**, 1–13 (2012).
2. A. Suppasri, N. Shuto, F. Imamura, S. Koshimura, E. Mas and A. C. Yalciner, Lessons learned from the 2011 Great East Japan Tsunami: Performance of Tsunami Counter-measures, Coastal Buildings, and Tsunami Evacuation in Japan, *Pure Appl. Geophys.* **170**, 993–1018 (2013).
3. E. Irtem, N. Gedik, M. S. Kabdasli and N. E. Yasa, Coastal forest effects on tsunami run-up heights, *Ocean Engineering.* **36**, 313–320 (2009).
4. N. Tanaka, effectiveness and limitations of vegetation bioshield in coast for tsunami disaster mitigagation. In: Mrner N-A (ed.) *The Tsunami Threat - Research and Technology*: InTech, 2011.

5. K. Iimura and N. Tanaka, Numerical simulation estimating effects of tree density distribution in coastal forest on tsunami mitigation, *Ocean Engineering.* **54**, 223–232 (2012).

6. W. Kellens, T. Terpstra and P. de Maeyer, Perception and communication of flood risks: A systematic review of empirical research, *Risk Anal.* **33**, 24–49 (2013).

7. Ghufran Ahmed Pasha and Norio Tanaka, Characteristics of a hydraulic jump formed on upstream vegetation of varying density and thickness, *Journal of Earthquake and Tsunami.* **14 (03)**, 2050012.

8. S. Matsuba, T. Mikami, R. Jayaratne, T. Shibayama and M. Esteban, Analysis of tsunami behavior and the effect of coastal forest in reducing tsunami force around the coastal dikes, *Int. Conf. Coastal. Eng.* **1**, 37 (2015).

9. F. Danielsen, M. K. Sørensen, M. F. Olwig, V. Selvam, F. Parish, N. D. Burgess, T. Hiraishi, V. M. Karunagaran, M. S. Rasmussen, L. B. Hansen, A. Quarto and N. Suryadiputra, The Asian tsunami: A protective role for coastal vegetation, *Science.* **310**, 643 (2005).

10. N. Tanaka, Y. Sasaki, M. I. M. Mowjood, K. B. S. N. Jinadasa and S. Homchuen, Coastal vegetation structures and their functions in tsunami protection: Experience of the recent Indian Ocean tsunami, *Landscape Ecol Eng.* **3**, 33–45 (2007).

11. N. Tanaka, S. Yasuda, K. Iimura and J. Yagisawa, Combined effects of coastal forest and sea embankment on reducing the washout region of houses in the Great East Japan tsunami, *Journal of Hydro-environment Research.* **8**, 270–280 (2014).

12. K. Harada1 and F. Imamura, Effects of coastal forest on tsunami hazard mitigation — A preliminary investigation. In: Satake K (ed.) *Tsunamis.* Berlin/Heidelberg: Springer-Verlag, 2005, pp. 279–292.

13. R. Ali Hasan Muhammad and N. Tanaka, Energy reduction of a tsunami current through a hybrid defense system comprising a sea embankment followed by a coastal forest, *Geosciences.* **9**, 247 (2019).

14. N. Tanaka, Vegetation bioshields for tsunami mitigation: Review of effectiveness, limitations, construction, and sustainable management, *Landscape Ecol Eng.* **5**, 71–79 (2009).

15. N. Tanaka, K. B. S. N. Jinadasa, M. I. M. Mowjood and M. S. M. Fasly, Coastal vegetation planting projects for tsunami disaster mitigation: Effectiveness evaluation of new establishments, *Landscape Ecol Eng.* **7**, 127–135 (2011).

16. Y. Igarashi and N. Tanaka, Effectiveness of a compound defense system of sea embankment and coastal forest against a tsunami, *Ocean Engineering.* **151**, 246–256 (2018).

17. L. Noarayanan, K. Murali and V. Sundar, Performance of flexible emergent vegetation in staggered configuration as a mitigation measure for extreme coastal disasters, *Nat Hazards.* **62**, 531–550 (2012).

18. N. Shuto, The effectiveness and limit of tsunami control forests, *Coastal Engineering in Japan.* **30**, 143–153 (1987).

19. K. Imai, H. Matsutomi, and T. Takahashi, Fluid force on vegetation due to tsunami flow on a sand spit, *Proceedings of Coastal Engineering.* **50**, 276–280 (2003).
20. N. Anjum and N. Tanaka, Experimental study on flow analysis and energy loss around discontinued vertically layered vegetation, *Environ Fluid Mech.* **20**, 791–817 (2020).
21. Afzal Ahmed and Abdul Razzaq Ghumman, Experimental investigation of flood energy dissipation by single and hybrid defense system.
22. A. Ahmed, M. Valyrakis, A. Razzaq Ghumman, G. A. Pasha and R. Farooq, Experimental investigation of flood energy dissipation through embankment followed by emergent vegetation, *Period. Polytech. Civil Eng.* (2020).
23. M. A. Rahman, N. Tanaka, A. H.M. Rashedunnabi and Y. Igarashi, Energy reduction in tsunami through a defense system comprising an embankment and vegetation on a mound, *In Proceedings of the 22nd IAHR-APD Congress.* (2020).
24. G. A. Pasha and N. Tanaka, Undular hydraulic jump formation and energy loss in a flow through emergent vegetation of varying thickness and density, *Ocean Engineering.* **141**, 308–325 (2017).
25. G. A. Pasha and N. Tanaka, Characteristics of a hydraulic jump formed on upstream vegetation of varying density and thickness, *J. Earthquake and Tsunami.* **14** (2020).
26. M. A. Rahman, N. Tanaka and A. H. M. Rashedunnabi, Flume experiments on flow analysis and energy reduction through a compound tsunami mitigation system with a seaward embankment and landward vegetation over a mound, *Geosciences.* **11**, 90 (2021).
27. G. A. Pasha, N. Tanaka, J. Yagisawa and F. N. Achmad, Tsunami mitigation by combination of coastal vegetation and a backward-facing step, *Coastal Engineering Journal.* **60**, 104–125 (2018).
28. R. de Costa and N. Tanaka, Role of hybrid structures on the control of tsunami induced large driftwood, *Coastal Engineering.* **163**, 103798 (2021).
29. T. Zaha, N. Tanaka and Y. Kimiwada, Flume experiments on optimal arrangement of hybrid defense system comprising an embankment, moat, and emergent vegetation to mitigate inundating tsunami current, *Ocean Engineering.* **173**, 45–57 (2019).
30. Y. Kimiwada, N. Tanaka and T. Zaha, Differences in effectiveness of a hybrid tsunami defense system comprising an embankment, moat, and forest in submerged, emergent, or combined conditions, *Ocean Engineering.* **208**, 107457 (2020).
31. N. Anjum and N. Tanaka, Investigating the effectiveness of discontinuous and layered coastal forest defense system against the inundating tsunami current, *Landscape Ecol Eng.* **18**, 171–190 (2022).
32. V. T. Chow. *Open-Channel Hydraulics*, International student edition. 21st ed.: Auckland: McGraw-Hill (McGraw-Hill civil engineering series), 1985.
33. N. Anjum and N. Tanaka, Hydrodynamics of longitudinally discontinuous, vertically double layered and partially covered rigid vegetation patches in open channel flow, *River Res Applic.* **36**, 115–127 (2020).

34. A. Rashedunnabi and N. Tanaka, Effectiveness of double-layer rigid vegetation in reducing the velocity and fluid force of a tsunami inundation behind the vegetation, *Ocean Engineering.* **201**, 107142 (2020).

35. S. Harish, B. Jochems, J. Oetjen, Holger Schüttrumpf, V. Sriram and S. A. Sannasiraj, Experimental investigation of tsunami bore momentum reduction using vegetation, *In Proceedings of the 23rd IAHR-APD Congress 2022, IIT Madras, India* (2023, in press).

36. N. Tanaka, J. Yagisawa and S. Yasuda, Breaking pattern and critical breaking condition of Japanese pine trees on coastal sand dunes in huge tsunami caused by Great East Japan Earthquake, *Nat Hazards.* **65**, 423–442 (2013).

37. N. B. Thuy, N. Tanaka and K. Tanimoto, Tsunami mitigation by coastal vegetation considering the effect of tree breaking, *J Coast Conserv.* **16**, 111–121 (2012).

38. N. Tanaka and Y. Igarashi, Multiple defense for tsunami inundation by two embankment system and prevention of oscillation by trees on embankment, *In Proceedings of the 20th congress of IAHR APD congress, Colombo, Sri Lanka*, 28–31 (2016, August).

39. Hokkaido Research Organization, Forestry and Forest Products Research Institute, and Saitama University, The Important Research Report on the Development of the Method to Control and Manage a Coastal Forest for Reducing Tsunami Energy (In Japanese). **2013.4-2015.3**, 69 (2016).

40. F. Usman, K. Murakami and E. B. Kurniawan, Study on reducing tsunami inundation energy by the modification of topography based on local wisdom, *Procedia Environmental Sciences.* **20**, 642–650 (2014).

41. S. Harish, V. Sriram, H. Schüttrumpf and S. A. Sannasiraj, Tsunami-like flow induced force on the structure: Prediction formulae for the horizontal force in quasi-steady flow phase, *Coastal Engineering.* **168**, 103938 (2021).

42. S. Harish, V. Sriram, H. Schüttrumpf and S. A. Sannasiraj, Tsunami-like flow-induced forces on the landward structure behind a vertical seawall with and without recurve using OpenFOAM, *Water.* **14**, 1986 (2022).

43. J. L. Lara, M. Maza, B. Ondiviela, J. Trinogga, I. J. Losada, T. J. Bouma and N. Gordejuela, Large-scale 3-D experiments of wave and current interaction with real vegetation. Part 1: Guidelines for physical modeling, *Coastal Engineering.* **107**, 70–83 (2016).

44. M. Maza, J. L. Lara and I. J. Losada, Experimental analysis of wave attenuation and drag forces in a realistic fringe Rhizophora mangrove forest, *Advances in Water Resources.* **131**, 103376 (2019).

45. T. Tomiczek, A. Wargula, P. Lomónaco, S. Goodwin, D. Cox, A. Kennedy and P. Lynett, Physical model investigation of mid-scale mangrove effects on flow hydrodynamics and pressures and loads in the built environment, *Coastal Engineering.* **162**, 103791 (2020).

46. S. A. Hughes. *Physical Models and Laboratory Techniques in Coastal Engineering*: World scientific, 1993.

47. H. Schüttrumpf and H. Oumeraci, Scale and model effects in crest level design, *In Proceedings of the 2nd Coastal Symposium, Höfn, Iceland*, 5–8 (2005).

48. N. Hari Ram, V. Sriram and K. Murali, Experimental investigation on the characteristics of solitary and elongated solitary waves passing over vegetation belt, *J. Ocean Eng. Mar. Energy.* **8**, 305–318 (2022).

49. G. Bullock, A. Crawford, P. Hewson, M. Walkden and P. Bird, The influence of air and scale on wave impact pressures, *Coastal Engineering.* **42**, 291–312 (2001).

50. M. Pfister and H. Chanson, Scale effects in physical hydraulic engineering models By VALENTIN HELLER, *Journal of Hydraulic Research*, Vol. 49, No. 3 (2011), pp. 293–306, *Journal of Hydraulic Research.* **50**, 244–246 (2012).

51. M. Pfister and H. Chanson, Two-phase air-water flows: Scale effects in physical modeling, *J. Hydrodyn.* **26**, 291–298 (2014).

52. V. Heller, Scale effects in physical hydraulic engineering models, *Journal of Hydraulic Research.* **49**, 293–306 (2011).

53. H. Chanson, Turbulent air–water flows in hydraulic structures: dynamic similarity and scale effects, *Environ Fluid Mech.* **9**, 125–142 (2009).

54. B. Le Méhauté. An introduction to water waves, In *An Introduction to Hydrodynamics and Water Waves*, Springer, Berlin, Heidelberg, 1976.

55. L. Noarayanan, K. Murali and V. Sundar, Manning's 'n' co-efficient for flexible emergent vegetation in tandem configuration, *Journal of Hydro-environment Research.* **6**, 51–62 (2012).

56. L. Noarayanan, K. Murali and V. Sundar, Manning's 'n' for staggered flexible emergent vegetation, *J. Earthquake and Tsunami.* **07**, 1250029 (2013).

57. K. Murali, V. Sundar and L. Noarayanan, Empirical equation for the prediction of run-up due to random waves on beaches fronted by vegetation, *Marine Geodesy.* **35**, 257–270 (2012).

58. S. Husrin, A. Strusińska and H. Oumeraci, Experimental study on tsunami attenuation by mangrove forest, *Earth Planet Sp.* **64**, 973–989 (2012).

59. J. Oetjen, V. Sundar, S. Venkatachalam, K. Reicherter, M. Engel, H. Schüttrumpf and S. A. Sannasiraj, A comprehensive review on structural tsunami countermeasures, *Nat Hazards.* **113**, 1419–1449 (2022).

Chapter 4

Experimental and Numerical Modelling of Flow-Vegetation Interaction in the Framework of Nature-Based Solutions (NBS) for Coastal Defense

Maria Maza

*IHCantabria – Instituto de Hidráulica Ambiental
de la Universidad de Cantabria,
Isabel Torres 15, 39011, Santander, Spain
mazame@unican.es*

Coastal areas are particularly vulnerable to climate change effects and are subjected to an increasing risk, mainly due to flooding and coastal erosion resulting from sea level rise and changes in the frequency and severity of extreme events. In addition, the rapid urbanization and subsidence of certain areas are increasing their exposure and vulnerability and, therefore, the need to manage these risks by adopting new adaptation measures. Among the most innovative strategies for coastal adaptation, are the nature-based solutions (NBS). NBS that include vegetated coastal ecosystems have demonstrated their ability to buffer incoming flow energy and retain sediment. However, the consideration of these NBS in coastal protection policies and adaptation plans is very limited. This is mainly due to the lack of guidelines and recommendations for their design, implementation and monitoring. These procedures should be based on the correct quantification of the coastal protection services provided by these solutions. This chapter presents a review of the different approaches followed at different scales to study the flow-vegetation interaction and to quantify the resulting ecosystem services in terms of coastal protection. The different ecological and hydrodynamic parameters involved at each scale and the obtained results are presented. Finally, an integrated approach incorporating the different scales is outlined and the main future needs to advance in the implementation of these NBS are identified.

1. Introduction

Coastal protection provided by natural ecosystems has been widely proven (e.g.: Shepard et al., 2011; McIvor et al., 2012a y b; Ferrario et al., 2014). This has led to specific actions to protect or restore natural ecosystems, simultaneously providing coastal protection, climate change adaptation and mitigation, biodiversity benefits and human well-being. These actions are known as Nature-Based Solutions (NBS) for coastal defense. Compared to traditional engineering solutions, NBS are of great interest due to their additional benefits on top of the provided disaster risk reduction (Morris et al., 2002; Duarte et al., 2013; Temmerman et al., 2013; Ondiviela et al., 2014). The main ecosystem services provided by NBS in terms of coastal protection are flood and coastal erosion risk reduction. That is, natural ecosystems attenuate hydrodynamic energy from tides and waves, and help to withhold sediment. In addition, they provide many other ecosystem services such as habitat creation or CO_2 capturing (Duarte et al., 2013). All these services make NBS an attractive solution compared to traditional engineering solutions. However, to firmly consider natural ecosystems in coastal protection policies, there is still a lack of tools and methodologies that help coastal managers to make decisions following clear steps and being able to quantify the protection provided and the associated risk.

Flooding and erosion risk reduction services depend on both, hydrodynamic and ecosystem properties (Figure 1). Traditionally when talking about coastal protection provided by ecosystems, main efforts have been done in understanding how ecosystems modify the flow and then attenuates its energy (e.g.: Dalrymple et al., 1986; Mendez and Losada 2004; Augustin et al., 2009) and, in a lesser extent, how morphodynamics are modified due to the presence of an ecosystem (e.g.: Tonelli et al., 2010; Zhou et al., 2014). Understanding the latest aspect is also extremely important when quantifying coastal protection provided by these ecosystems. Moreover, flow conditions and ecosystems interact with each other, which may lead to different dynamic equilibriums that change under variations of any of both agents. Understanding the conditional outcome of this interaction requires of more attention, since ecosystem services strongly relies on it. Future protection will be conditioned by ecosystem adaptation to the new flow conditions and their resilience. Then, engineers and ecologists should work together to identify the different agents involved in the problem, and their interactions, to be able to develop integrated tools and methodologies. This implies interdisciplinary integration and cooperation, starting from integrate the disparate vocabularies, basic concepts, measured variables and approaches followed by engineers and ecologists to continue defining new terms and variables and developing an interdisciplinary working methodology. This integrated approach will help including natural ecosystems in coastal protection policies.

At the same time, this interdisciplinary methodology should be based on fundamental understanding and basic research principles to help setting new parameterizations and formulations in a synthetic, organized and integrated manner (Nestler et al. 2016). To be able to address real problems at big scales, the physics and biological parameters involved at the small-scale need to be simplified. The simplifications and assumptions considered to quantify coastal protection provided by an ecosystem at a regional or global scale should be sustained by a deep understanding of the small-scale phenomena that are not reproduced, but instead they are disregarded or parameterized, at a large scale (Figure 1). Join to this, spatial and temporal scale variations, including both hydrodynamic and ecological changes, should be included in the integrated methodology to be able to give predictions of the future vulnerability of the protected area and the associated risk.

Despite all the above-mentioned needs, in the last decades, there have been a growing number of studies that have resulted in a deeper understanding of the flow-ecosystem interaction, which provide the scientific basis for developing tools and establishing methodologies to firmly embed NBS in the decision-making process of coastal managers. This interaction is extremely complex due to the complex hydrodynamics and ecosystem properties involved. Coastal ecosystems grow in coastal areas either at the intertidal zone or submerged. The characteristics of the organisms compounding these ecosystems widely vary from emergent and rigid individuals to submerged highly flexible ones. The main coastal vegetated ecosystems that serve with coastal protection services are: seagrasses, saltmarshes and mangroves. Mangrove forests develop in intertidal areas in tropical zones. At these areas, but in temperate zones, saltmarshes are present. Seagrasses grow in shallow water environments at both, temperate

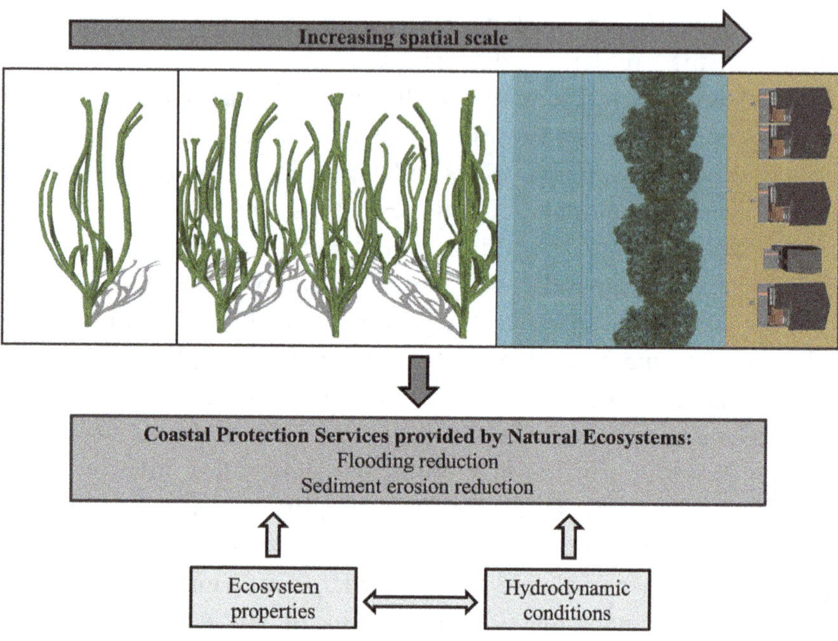

Figure 1. Conceptual sketch of the coastal protection services provided by vegetated ecosystems that are influenced by both, ecosystem properties and hydrodynamic conditions. Increasing spatial scale these services are estimated as a function of the interaction of the flow with individual ecosystem elements till the consideration of the whole ecosystem or several ecosystems.

and tropical zones. Then, flow-vegetation interaction differs depending on the studied ecosystem, which leads to strongly different coastal protection services. Therefore, the protection provided by each community should be studied in terms of its properties and the hydrodynamic conditions at which it is subjected.

In this chapter, an overview of flow-vegetation interaction modeling in the framework of NBS for coastal defense is presented. The different scales and main ecological and hydrodynamic parameters at each scale are identified. Studies ranging from the individual organism scale to the ecosystem or several ecosystems scale, including studies at the community scale (sketch in upper panel in Figure 1) are presented in sections 2, 3 and 4, including numerical and experimental approaches at the different scales.

Finally, section 5 presents how the different scales can be linked in an integrated approach and details the most important future needs.

2. Individual organism scale

At the small scale, vegetated coastal ecosystems present different properties. Morphologies of organisms are, in general, very complex presenting networks conformed by leaves, branches, trunks, roots or rhizomes. In addition to this, they present different biomechanical properties, from organisms that behave as rigid bodies under the flow action presenting a fragile failure, as mature mangrove trees, till some highly flexible ones that move passively under the flow action, as some seagrasses. Both, morphology and biomechanical properties directly influence the organism behavior under the flow action and determine the associated turbulent processes and the forces exerted on the organisms by the flow.

Numerical and experimental studies have been performed to characterized flow interaction with individual organisms. These studies have been mainly focused on analyzing turbulence processes and the forces exerted on the organism by the flow. Numerical models at this scale, solve Navier-Stokes equations using different approaches ranging from the Reynolds Averaged Navier Stokes equations (RANS) and Large Eddy Simulations (LES), that use turbulence closure models, till Direct Numerical Simulations (DNS), that do not use turbulence closure models, but instead solve all temporal and spatial scales. However, the use of these last ones in studying flow-vegetation interaction is still open.

Laboratory experiments have been also conducted to well characterize the small-scale processes involved in flow-vegetation interactions. These studies are usually conducted using laser techniques such as Particle Image Velocimetry (PIV), Laser Doppler Velocimetry (LDV) or Laser Doppler Anemometry (LDA). Also, special attention has been paid to the forces exerted on the organism by the fluid using devices, such as load cells or load plates, to quantify this force. In the following, a state of the art of different studies using these models and techniques is performed based on the processes that are studied. Then, the main findings at this scale in terms of the coastal protection services provided by NBS are highlighted below.

Turbulent processes have been studied analyzing individual plants or small canopies conform by few individuals. Processes, such as turbulent intensity, diffusion or turbulence scales, have been investigated for several years providing analytical models that have been tested using laboratory experiments (Nepf, 1999; Lopez and García, 2001; King et al., 2012; Abdolahpour et al., 2018). Coherent structures have also been studied, including vortex shedding from individual elements using laser techniques in the laboratory (Ghisalberti and Nepf, 2002) or running numerical simulations solving LES equations (Cui and Neary, 2002; Stoesser et al., 2009, 2010). As an example, left panel in Figure 2 shows the flow separation produced around an idealized vegetation stem and the updraft produced behind it. Also, some other coherent structures, such as monamis produced on top of a submerged flexible meadow under unidirectional flow have been studied in the laboratory using PIV (Okamoto et al., 2016) or solving RANS equations (Singh et al., 2015). Most of these studies have been performed using simplified geometries, usually rigid cylinders, and studies using simplified flexible mimics are scarce. Few studies have been focused on characterizing turbulent processes using real organisms (e.g.: Shucksmith et al., 2010; Sonnenwald et al., 2017) overcoming the problem of correctly representing their properties. In addition, most of the studies have been performed considering unidirectional flow conditions that usually are not related, neither scaled, to real environments. Despite of these limitations, these studies have allowed getting a deeper understanding on the small-scale processes in the flow-vegetation interaction. Additionally, they have led to new parameterizations or formulas than can be used in larger scale models, such as the modified k-epsilon model for canopy flow presented by Hiraoka et al. (2006) that has been implemented in RANS models (e.g: Maza et al., 2013).

Other important process influenced by turbulent structures and intensity is the developed boundary layer next to the bottom, and the resultant shear stresses. These shear stresses are especially important when analyzing sediment transport in vegetated environments. Bottom shear stresses around individual elements determine the possible scour around the organisms, which can lead to their uprooting (e.g.: types of mangrove failures described in Yanagisawa et al., 2009). Bottom shear stresses can be evaluated in different ways. For example, they are usually evaluated by fitting the Log Law of the Wall to get the mean velocity profile near the bed. However, this does not work in vegetated environments where mean velocity profile is not logarithmic (Nezu and Nakagawa, 1993;

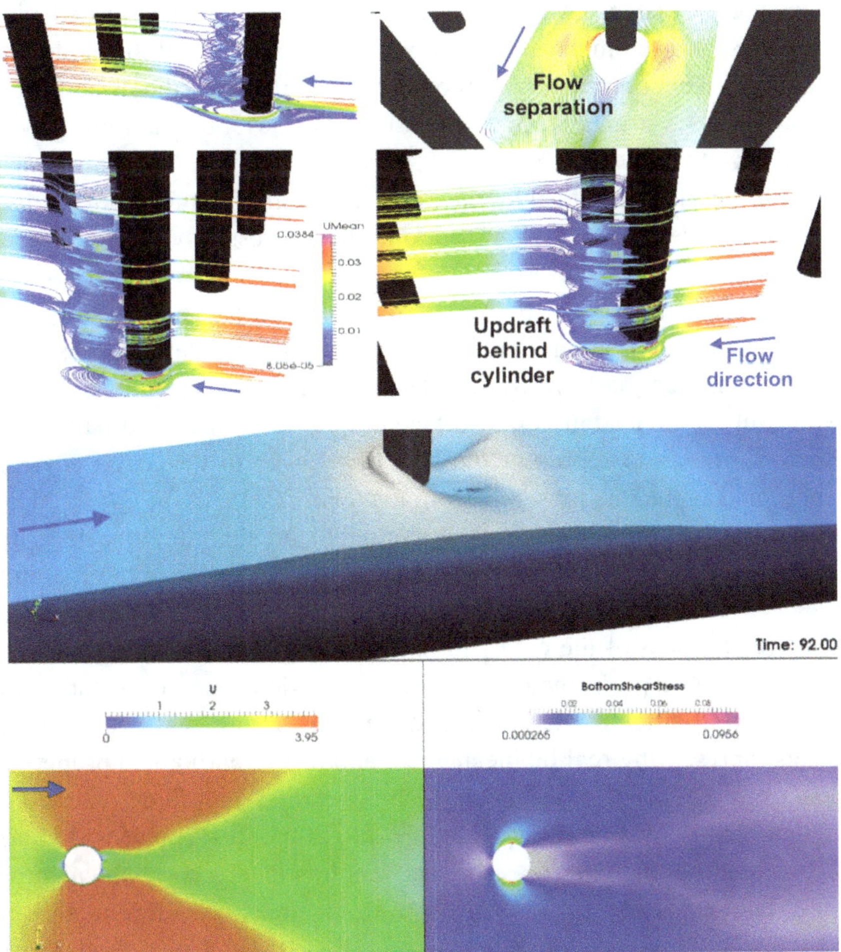

Figure 2. Visualization of two simulations performed by using IHFOAM model to show flow separation and bottom shear stress, blue arrows show flow direction. Upper panel: example of streamlines colored by mean velocity magnitude for a unidirectional flow test interacting with a cylinder (flow separation and updraft on the back face of the cylinder are displayed). Bottom panel: interaction of regular waves flowing with a following current with a cylinder (free surface visualization, velocity at half water depth, U, and bottom shear stress are displayed).

Liu et al., 2008). Different studies have been performed to better understand this process in vegetated environments in the laboratory (e.g.: Zong and Nepf, 2010; Montakhab et al., 2012) and numerically using RANS and LES equations (e.g.: Li and Yu, 2010; Kim and Stoesser, 2011). Again, these studies have been developed considering simplified plant geometries and biomechanical properties and have not been defined attending to real conditions. Additionally, the number of studies performed at this small scale under waves or waves and current conditions is very limited. Right panel in Figure 2 shows an example of a numerical simulation performed considering waves and currents interacting with an isolated cylinder. The resultant bottom shear stress when a wave crest interacts with the cylinder is shown for this simulation run using a RANS model, displaying the complex pattern developed around the simplified element. Despite these limitations linked to the simplifications considered in both numerical and experimental studies, they have allowed moving forward in the understanding of these complex processes and have result in new formulas to represent the boundary layer in vegetated environments, such as the one presented by Yang et al., 2015.

Another parameter that has been studied on this small-scale is the force exerted by the flow on the organism. Many studies in the literature have been done to characterize this force for different types of organisms subjected, in most of the cases, to unidirectional flow. Although several studies have been performed considering simplified geometries, especially cylinders, less attempts have been done considering real organisms or mimics representing real biomechanical and/or morphological properties. Some examples are: Callaghan et al., 2007 studied drag forces exerted on real and plastic macrophytes; Bouma et al., 2005, 2010 presented values for two real saltmarshes species with contrasting biomechanical properties; Jalonen et al., 2013 studied the drag forces on different flexible vegetation mimics; Strusinska-Correia et al., 2013 did it in model mangroves, built using rigid cylinders and, in that case, testing solitary waves and bores; Chapman et al., 2015 did a comparison of drag forces exerted on rigid and flexible elements; Maza et al. 2017 studied the drag forces exerted on a 3-D Rhizophora mangrove model built considering its prompt roots; and Maza et al. 2019 studied the drag forces on 3-D Rhizophora models but, in this case, under the action of waves. Obtained forces are usually associated only to drag force component, disregarding other forces such as the inertia force or buoyancy force that can be especially important for flexible individuals. This simplification considering only drag force, and then only one calibration coefficient, the drag coefficient, can be useful to parameterize the problem at the small-scale and then use the calibrated coefficient at bigger scales. However, the importance of additional forces in each specific problem should be addressed and for cases when other forces are important, they should be considered (e.g.: Gosselin et al., 2010; Dijkstra and Uittenbogaard, 2010; Luhar and Nepf, 2010 and 2011). This drag force is formulated as:

$$F_D = \frac{1}{2}\rho C_D A_f u|u| \tag{1}$$

where ρ is the fluid density, A_f is the organism frontal area, u is the flow velocity and C_D is the drag coefficient. As can be observed, the formulation of this force does not directly include neither real morphology nor bio-mechanical properties of the plant. However, both strongly influence the force experienced by the organism. Then, these missing parameters, are included indirectly in the value used for C_D. This C_D is used to calibrate the force exerted over each organism. That is, since variables defining biomechanical properties and complex morphologies are not included in the equation, C_D serves as a calibration parameter to account for those missing variables. However, there is not a formulation that relates C_D to organism's properties and then, a calibration of this coefficient is needed for each case. Looking for a parameterization of this coefficient, many studies in the literature have related the obtained drag coefficients to the Reynolds number, Re, or the Keulegan-Carpenter number, KC, defined as:

$$Re = \frac{uL}{\nu} \quad ; \quad KC = \frac{uT}{L} \tag{2}$$

where L is the characteristic length of the individual, usually taken as its width and for cases where cylinders are used their diameter, ν is the kinematic viscosity of the fluid and T is the period of the oscillation, using the wave period for cases subjected to waves action. As can be observed, these non-dimensional numbers neither include the individuals' properties further than considering a characteristic length scale. However, although they are case specific, the set of relationships between C_D and these non-dimensional numbers found in the literature, allow getting an estimation of this coefficient for different flow and plant properties.

Previous literature review has been focused on the effects of plants on the flow. However, these turbulent processes and forces also directly influence the organisms. Previous studies have shown that vegetation properties can change in response to the local hydrodynamics (e.g.: Puijalon et al., 2005; Peralta et al., 2006; Stewart, 2006). This shows the strong feedbacks between flow and organisms, and associated processes such as nutrient and oxygen exchange or organism survival. In the framework of the application of NBS for coastal protection, it is crucial to obtain a relationship between the forces exerted by the flow and the

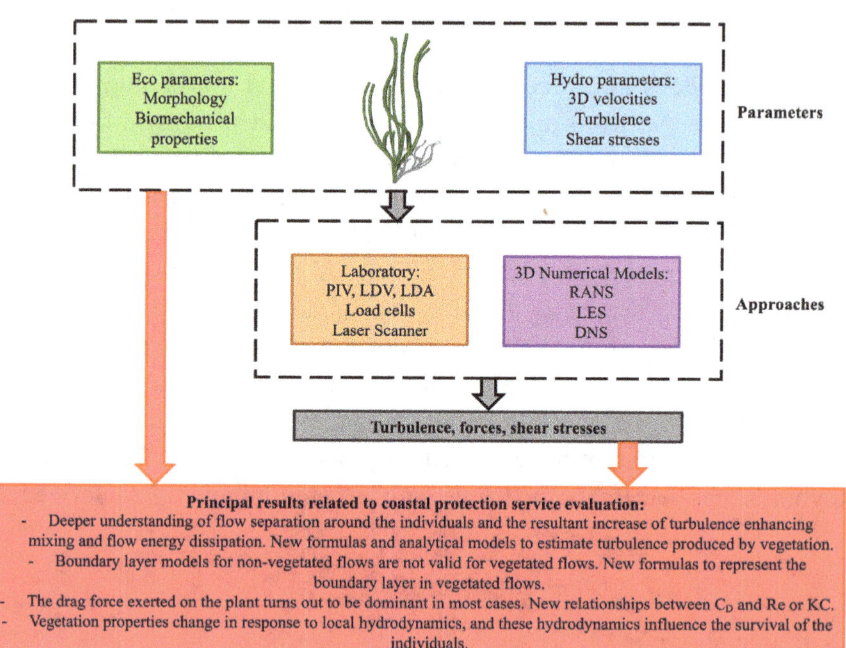

Figure 3. Framework for individual organism scale studies: ecological (green) and hydrodynamic (blue) parameters and different approaches (experimental in orange and numerical in purple) including main techniques and equations. Outputs obtained from these approaches are presented in grey and the principal results related to the assessment of the coastal protection service provided by vegetated ecosystems are presented in red.

turbulence around the organisms and their potential uprooting or break-up to assess the associated risk. Although a huge effort is still needed linking hydrodynamics and ecology looking not only for the effect of the individuals on the flow, but also for the implications that this interaction has on the individuals, existing studies have set the basis to understand this double interaction.

Figure 3 summarizes the different hydrodynamic and ecological parameters considered at this individual scale and the output obtained from the different approaches, as well as the main aspects studied at this scale related to the assessment of the coastal protection service provided by vegetated ecosystems.

3. Community scale

Coastal vegetated communities are formed by diverse individuals distributed following different patterns leading in many cases to non-continuous fields or changes in the community properties along its extension. These changes can be linked to different densities (number of individuals per square meter) or changes in the individuals' properties (e.g.: different morphology or biomechanical properties). These changes determine the influence of the community in the flow properties.

Studies at the meso-scale, considering communities conformed by several individuals, are the most common in the literature when analyzing coastal protection services. In these studies, small scale processes such as turbulent structures or boundary layer shape, are not studied in detailed or resolved, but instead these studies are more focus on quantifying the bulk energy attenuation produced by the community or, in a lesser extent, analyze the transport processes produced at and around the community. Studies based on experimental modeling, commonly use free surface gauges and Acoustic Doppler Velocimeters (ADV) to quantify the flow energy attenuation produced by an ecosystem. To measure sediment changes, techniques range from yardsticks to sophisticated laser scanners. Numerical approaches are usually based on Non-Linear Shallow Water equations (NLSW), Boussinesq equations, and, in more detailed studies, RANS.

M. Maza

Many studies have been performed to quantify flow energy attenuation produced by vegetated communities. In the laboratory, studies performed considering unidirectional flow (e.g.: Nepf, 1999; Bouma et al., 2013) have quantified flow resistance and current velocity attenuation produced by a vegetation field by measuring velocities and free surface at different positions. These studies establish a momentum balance between the drag force and the free surface gradient produced along the vegetation field, as follows:

$$gh\frac{\partial h}{\partial x} = \frac{1}{2}C_D aNhU_0^2 \tag{3}$$

where g is the acceleration of gravity, h is the flow depth, x is the longitudinal distance along the meadow, N is the number of individuals per unit horizontal area, a is the organism's mean width and U_0 is the mean uniform velocity.

Experiments run considering wave conditions (e.g., Bouma et al., 2010; Möller et al., 2014; Maza et al., 2015a) have measured wave height decay along the community to later apply an energy balance considering that, the energy dissipation is only due to the drag force produced by the vegetation. Dalrymple et al. (1984) first derived this relationship assuming linear wave theory and assuming a constant wave damping coefficient over depth. Thus, the wave height decay, and the associated drag coefficient, were expressed as:

$$H = \frac{H_0}{1 + \beta x} \quad ; \quad C_D = \frac{9\pi}{4aNH_0 k}\frac{(sinh2kh + 2kh)sinhkh}{sinh^3 kl + 3sinhkl}\beta \tag{4}$$

where H is the wave height and β the wave damping coefficient for regular wave trains, the subscript 0 refers to the incident wave height, k is the wave number, h is the water depth and l is the organism's height. Mendez and Losada (2004) extended this formulation for nonbreaking random waves over constant depth assuming a Rayleigh distribution and deriving the wave height evolution as a function of the root-mean-square wave height as:

$$H_{rms} = \frac{H_{rms,0}}{1 + \beta' x} \quad ; \quad C_D = \frac{3\sqrt{\pi}}{aN H_{rms,0} k} \frac{(sinh2kh + 2kh)sinhkh}{sinh^3 kl + 3sinhkl} \beta' \quad (5)$$

where H_{rms} is the root-mean-square wave height and β' the wave damping coefficient for random wave trains. Several studies have fit their experimental data to these formulations getting the associated wave damping coefficients and the drag coefficient (e.g.: Koftis et al., 2013; Ozeren et al., 2013; Anderson et al., 2014). A recent study, Maza et al. 2019, also based on the fitting of these formulas to their experimental data, highlighted the importance of considering the friction produced by the bottom and walls of the experimental facility when obtaining β and C_D for a vegetation field. This friction also produces wave attenuation and, if it is not considered in the analysis, the wave attenuation capacity of the ecosystem can be significantly overestimated, especially for cases with long vegetation fields and narrow flumes.

Although few studies have been performed combining waves and currents (e.g.: Gaylord and Denny, 2003; Paul et al., 2012; Hu et al., 2014; Maza et al., 2015a), this energy balance was also extended to waves and current conditions by Losada et al. (2016), for regular waves:

$$H = \frac{H_0}{1 + \beta_{wc} x} \quad ;$$

$$C_{Dwc} = \frac{3\pi}{2aN \left(\frac{gk}{2(\sigma - U_0 k)}\right)^3 H_0} \frac{3kcosh^3 kh}{sinh^3 kl_D + 3sinhkl_D} \left[\frac{g}{8} \left(1 \right. \right.$$

$$\left. + \frac{2kh}{sinh2kh} \right) \left(\frac{g}{k} tanhkh\right)^{\frac{1}{2}} + \frac{g}{8} U_0 \left(3 + \frac{4kh}{sinh2kh} \right)$$

$$\left. + \frac{3k}{8} U_0^2 \left(\frac{g}{k} cothkh\right)^{\frac{1}{2}} \right] \left[U_0 \right.$$

$$\left. + \frac{1}{2} \left(1 + \frac{2kh}{sinh2kh} \right) \left(\frac{g}{k} tanhkh\right)^{\frac{1}{2}} \right] \beta_{wc} \quad (6)$$

where β_{wc} and C_{Dwc} are the wave damping coefficient and drag coefficient for regular waves and currents flowing in the same and in opposite direction to wave propagation, U_0 is the uniform current velocity and l_D is the deflected length of the plants under the flow action, which is equal to the length of the plants for plants considered rigid under the flow action. They also derived the expression for random waves:

$$H_{rms} = \frac{H_{rms,0}}{1 + \beta'_{wc} x} \; ;$$

$$C'_{Dwc} = \frac{2\sqrt{\pi}}{aN \left(\frac{gk}{2(\sigma - U_0 k)}\right)^3 H_{rms,0}} \frac{3k cosh^3 kh}{sinh^3 kl_D + 3 sinh kl_D} \left[\frac{g}{8}\left(1 + \frac{2kh}{sinh2kh}\right)\left(\frac{g}{k} tanhkh\right)^{\frac{1}{2}} + \frac{g}{8} U_0 \left(3 + \frac{4kh}{sinh2kh}\right) + \frac{3k}{8} U_0^2 \left(\frac{g}{k} cothkh\right)^{\frac{1}{2}}\right]\left[U_0 + \frac{1}{2}\left(1 + \frac{2kh}{sinh2kh}\right)\left(\frac{g}{k} tanhkh\right)^{\frac{1}{2}}\right]\beta'_{wc} \tag{7}$$

where β'_{wc} and C'_{Dwc} are the wave damping coefficient and drag coefficient for random waves and currents flowing in the same and in opposite direction to wave propagation. Looking for parameterizations that allow getting the value of C_D for different cases, C_D values as commonly presented as a function of Re and KC numbers. These C_D values are obtained based on the relationship between the energy attenuation produced by the vegetation field and the drag force shown in previous equations. Then, they can be considered as bulk coefficients that allow the estimation of the energy attenuation produced by the ecosystem.

Studies based on these energy balances, as the ones presented in this section, represent the community by using the geometrical variables: N, a and l. However, it is important to keep in mind that natural communities present complex geometries both in the horizontal dimension (patchiness and different densities) and vertically (non-uniform distribution of plant geometry and different submergence ratios, defined as the ration between the plant height and the water depth) and they can be formed by individuals with contrasting biomechanical properties. Several studies have been developed to further study the influence of these parameters in the induced flow energy attenuation. Some examples are Paul et al. 2012, where different seagrass mimics were studied to understand the influence of blade stiffness, length and shoot density; Bouma et al. 2015, where three contrasting marsh species were tested in the laboratory under different current velocities identifying different flow patterns depending on the plants characteristics and; Maza et al. 2016, where different patches of mangrove roots were tested under different solitary waves analyzing wave diffraction and attenuation depending on the patches distribution (panel (a) in Figure 4 shows the case of one patch). To account for the influence of the patches' distribution in the estimation of wave height attenuation, they proposed an equivalent length, that is obtained based on patches geo-metrical properties and results in the length of a uniform vegetation field inducing the same wave height attenuation.

As shown in this section, most of the studies found in the literature have been based on the consideration of a drag force and different values of C_D have been obtained for different types of vegetation fields and flow conditions, being case dependent. Although a huge effort has been done in the characterization of C_D, this has led to a large number of formulas for estimating its value. Considering this, and the complex geometry and biomechanical properties of real communities, new approaches have been investigated to quantify the induced energy attenuation. One of them, has been focused on the consideration of the submerged solid volume fraction, SVF, of communities that can be considered rigid under the flow action, such as mature mangrove forests. Mazda et al. (1997) already reported the importance of the SVF in the resultant wave attenuation produced by mangroves. Horstman et al. (2014) also quantified this SVF and high-lighted the increasing wave attenuation obtained for larger SVF values. More recently, Maza et al. (2019) found a direct relationship between the wave damping coefficient and this SVF as follows:

$$\beta = 1.295 \frac{H}{h} SVF + 0.004 \tag{8}$$

$$\beta' = 1.783 \frac{H_{rms,i}}{h} SVF + 0.001 \tag{9}$$

where β and β' are the wave damping coefficients for regular and random waves and *SVF* is obtained as the ratio of the submerged forest volume and the water volume. These relationships allow getting an estimate of the wave damping coefficient by quantifying the *SVF* and without the need of calibrating C_D.

Another approach, based on the ecosystem standing biomass, has been investigated for communities of different geometries and biomechanical properties. The standing biomass is defined as the plants weight per unit area (gr/m^2). Bouma et al. (2010) found a linear relationship between this parameter and the obtained wave height attenuation for two different saltmarsh vegetation species. Maza et al. (2015a) also found a linear relationship between the wave damping coefficient and the standing biomass for two saltmarsh species tested considering different densities, including cases run under the combined effect of waves and currents (panel (b) in Figure 4). In these studies, the linear relationship between the wave damping coefficient and the vegetation field standing biomass was confirmed, but a different linear relationship was found for each tested wave condition. To further explore this promising finding and looking for a common relationship for different wave conditions, Maza et al. (2022) tested 4 different saltmarsh species in the laboratory, considering two densities per species, which resulted in eight standing biomass values. A wide range of waves representative of daily conditions affecting the selected species were run. The linear relationship between the wave damping coefficient and the standing biomass was confirmed and they defined a new parameter, the hydraulic standing biomass, *HSB*, that includes not only the ecosystem standing biomass and mean ecosystem height, but also the incident wave height and period and the water depth to account for different wave conditions affecting the ecosystem. By considering this parameter, a unique linear relationship was obtained for all standing biomass values and the full range of wave conditions. They identified a saturation regime, in which the wave height beyond the meadow can be assumed to be negligible leading to a two-section fitting relationship. The equations found for regular and random waves are shown in equations 10 and 11.

$$\beta = \begin{cases} 1.020 \cdot 10^{-3} * HSB + 0.088 & 0 < HSB < 659 \\ 0.758 & HSB > 659 \end{cases} \quad (10)$$

$$\beta' = \begin{cases} 1.310 \cdot 10^{-3} * HSB + 0.059 & 0 < HSB < 474 \\ 0.684 & HSB > 474 \end{cases} \quad (11)$$

Maza et al. (2022) also accounted for the dissipation produced by bottom friction by running tests where all vegetation was removed and the wave height attenuation produced by the bare soil was recorded. This results in two additional equations that allow accounting for the wave damping produced solely by the vegetation standing biomass. The equations for regular and random waves are given in equations 12 and 13, respectively, where β_{SB} and β'_{SB} are the wave damping coefficients obtained after subtracting the bottom friction contribution.

$$\beta_{SB} = \begin{cases} 1.151 \cdot 10^{-3} * HSB & 0 < HSB < 599 \\ 0.685 & HSB > 599 \end{cases} \quad (12)$$

$$\beta'_{SB} = \begin{cases} 1.396 \cdot 10^{-3} * HSB & 0 < HSB < 451 \\ 0.631 & HSB > 451 \end{cases} \quad (13)$$

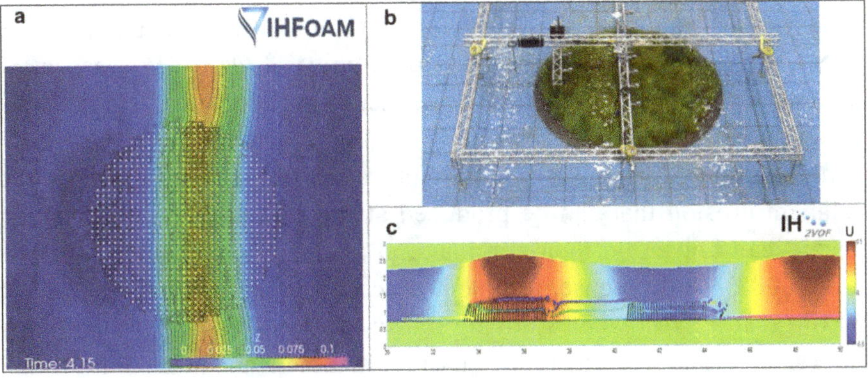

Figure 4. Examples of studies performed at the community scale: (a) wave height attenuation by a vegetation patch (Maza et al. 2016), (b) interaction of waves and currents with real saltmarshes (Maza et al. 2015a) (c) theoretical sediments and nutrients transport inside and around vegetation patches (Maza et al. 2013b).

This new approach for estimating the wave damping coefficient is a breakthrough in the estimation of wave attenuation by different vegetation types, as it depends on the standing biomass, a variable widely known to ecologists and that can be obtained from aerial imagery and remote sensing, and a mean meadow height, avoiding the use of any calibration coefficient.

Meso-scale numerical models, which are able to simulate an entire community, i.e., domains from tens of meters to a few kilometers, account for parameterizations of the vegetation since they are not capable of simulating the small-scale physics involved in the flow-vegetation interaction due to the inherent limitations of the flow equations they resolve or the associated computational effort. These models are based on the implementation of a drag force to resolve the flow-vegetation interaction and infer the energy attenuation produced by the vegetation field. Some examples are the full spectrum model SWAN (Suzuki et al., 2012), including the wave damping model by Mendez and Losada (2004), or Chen et al. (2007) considering the effects of seagrass bed geometry on wave attenuation and suspended sediment transport using a modified Nearshore Community Model (NearCoM). Examples of phase resolving models able to solve the kinematics and dynamics within the vegetation field are the Boussinesq equation-based model developed by Augustin et al., (2009), the 2-D RANS model IH2VOF (Maza et al., 2013a, panel (c) in Figure 4) and the 3-D RANS model IHFOAM (Maza et al., 2015b).

Sediment transport around the community is also a key aspect to account for the coastal protection service considering the erosion risk reduction provided by the community and its potential for soil establishment. Additionally, another important aspect to be considered is the sediment erosion that can be produced at the community and its edges, since it can impact on their survival and lateral development (Widdows et al., 2008). In this sense, turbulent processes studied at the small-scale are of great interest when studying transport processes at the community scale since local turbulence can determine sediment and nutrient transport, being even more important than mean dynamics affecting the individuals. An example of this effect is observed when comparing two canopies subjected to the same unidirectional flow but with different densities (Nepf, 2012; Nardin and Edmonds, 2014). A sparse canopy can produce erosion in the area occupied by the vegetation due to the turbulent structures produced between the individual elements, as was shown in the experiments presented by Bouma et al. (2009). On the contrary, a dense canopy subjected to the same hydrodynamic conditions can result into sedimentation inside the canopy since flow velocity is strongly reduced and turbulent structures between elements are broken due to the proximity between the individuals (Bouma et al., 2009). However, if the canopy is very dense, suspended sediment does not penetrate it and sedimentation inside the canopy is not produced. Then, turbulence at the small and the community scale determine the sediment transport inside and around the community. Although, there is still a lack of formulations that relate these processes to real communities' characteristics, studies performed with canopies made of rigid cylinders (e.g.: Zong and Nepf, 2010 and 2011) and rigid and flexible mimics (Gillis et al., 2022) have provided with formulations based on the canopy density and solid volume fraction that provide the basis for formulations for more complex geometries. Additionally, tests performed using real vegetation (e.g.: Gambi et al., 1990; Bouma et al., 2009) have given results for the specific cases under study.

In addition to the role played by turbulence, flow velocities and local shear stresses also determine these transport processes. Local accelerations or decelerations result in sediment erosion or accretion (see example at Figure 4c, where passive particles are transport in a submerged flexible canopy under the waves action: particles are transported onshore on top of the canopy due to the strong shear layer present at that location, and particles behind the two vegetation patches are recirculated due to the resultant local flow deceleration behind each patch). Then, all these processes drive morphological changes in and around the community (Larsen and Harvey, 2010), specially at community edges (Fagherazzi et al., 2013; Gillis et al., 2022). Additionally, transport processes determine ecosystem health since they are responsible of nutrient transport inside the ecosystem and light availability, e.g. an area full of suspended sediment does not represent a good environment for healthy submerged ecosystems. Furthermore, these processes will be determinant in the ability of the ecosystem to colonize adjacent areas and expand, e.g. leading and lateral edges of the ecosystem that experience sediment erosion will limit the expansion of the ecosystem. These implications of hydrodynamic processes on biological factors are also very relevant in the consideration of NBS implementation as they will influence the survival of the ecosystem and its potential growth.

Numerical models solving sediment transport are usually based on van Rijn (1984) and Soulsby (1997) formulations. Ma (2014) solve RANS equations including the effect of vegetation in the momentum equation and the turbulence closure model but following the Law of the Wall at vegetated areas which is usually not met, as discussed in previous section. Roelvink et al. (2009) solve NLSW equations getting the critical velocity for sediment transport initiation considering a Shields parameter and a bottom friction coefficient (i.e. Manning coefficient). This coefficient has been enhanced by some authors to include the vegetation effect (e.g.: Wu et al., 2005; Chen et al., 2007). The non-hydrostatic mode of X-Beach (McCall et al., 2014) model, originally developed as a phase-averaged model, allows to fully resolve sea-swell waves and includes sediment transport and morphological changes by choosing the sediment transport formulations (such as Soulsby (1997) and van Rijn (1984) equations). The model has been extended to include the effect of wave-vegetation interaction by van Rooijen et al. (2016), who presented an application to study wave setup. Although these models can be applied to study sediment transport processes in vegetation fields, they still involve strong simplifications and depend on coefficients that need to be calibrated for each case study. Therefore, further development is still needed, considering the efforts made at the micro-scale, to come up with tools to adequately estimate the coastal protection service provided by vegetated ecosystems in terms of erosion risk reduction.

Figure 5 shows a sketch of the different parameters, ecological and hydrodynamic, that are considered in the different approaches at the meso-scale. The analysis of the results obtained with these approaches, experimental and numerical, have allowed identifying the outcomes related to the coastal protection service evaluation.

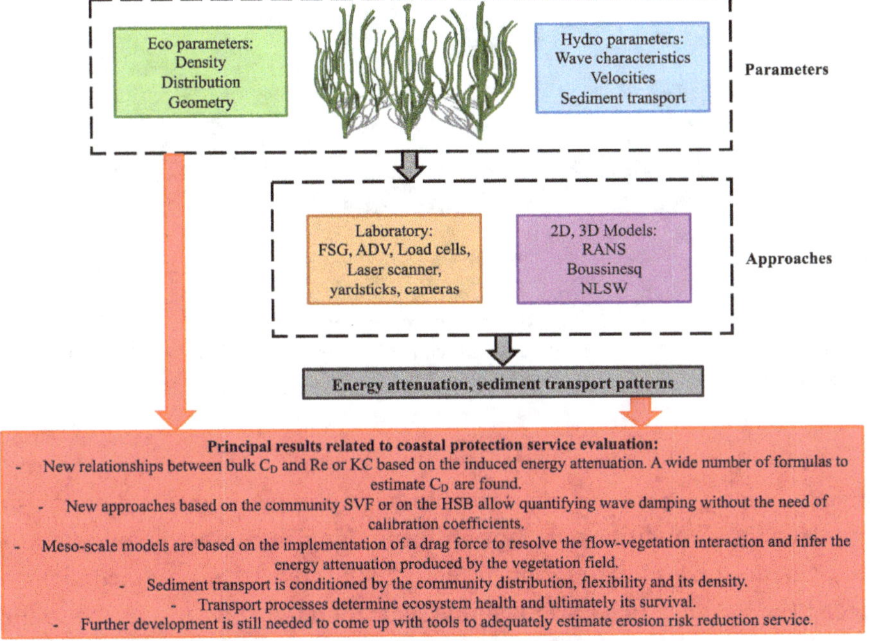

Figure 5. Framework for community scale studies: ecological (green) and hydrodynamic (blue) parameters and different approaches (experimental in orange and numerical in purple) including main techniques and equations. Outputs obtained from these approaches are presented in grey and the principal results related to the assessment of the coastal protection service provided by vegetated ecosystems are presented in red.

4. Ecosystem or several ecosystems scale

In the last few years, large scale models have become more and more necessary since the integration of NBS in coastal protection policies and climate change adaptation plans has started to become a reality. At this scale, we found complete ecosystems and even more than one ecosystem interacting with the flow and sediment in the area and then providing with coastal protection services. When several ecosystems are present in the area, they should be all studied since they influence each other flow energy attenuation capacity, as illustrated in Figure 6.

Figure 6. Sketch of wave interaction with an area where seagrasses and mangroves are present: the interaction between ecosystems should be considered, i.e. wave height attenuation by the seagrass meadow influences the incident wave height affecting the mangrove forest and the resultant final coastal protection service provided by both ecosystems.

At this large scale, ecosystems are not continuous, and their properties change along the ecosystem including different communities. Also, incident flow conditions are not uniform along the entered ecosystem or ecosystems. These variations directly determine the provided coastal protection service. In addition, at this scale long time periods, from months to years, should be considered if coastal protection service wants to be analyzed and quantified, especially for ecosystems that significantly change their properties seasonally or with age and areas where climate change effects are expected to be important. Along the different seasons, flow conditions can vary significantly, e.g. in temperate zones storms are commonly stronger during the winter whereas tropical cyclones are more probable during the autumn. In addition, many ecosystems, e.g. salt-marshes species such as *Spartina*, dramatically change their properties, in terms of frontal area and standing biomass, along the year. Then, ecosystem properties variability should be considered when quantifying coastal protection services. If this variability is not included, the ecosystem protection capacity could be overestimated. For example, saltmarshes present the smallest density, height and standing biomass during the winter, when the strongest storms hit temperate zones. Then, the eco-system exhibits its smallest attenuation capacity when worst flow conditions get the coast. This scenario should be considered as the one resulting in the highest risk to the protected area. However, although several studies have been performed to quantify the seasonal variability of different ecosystems (e.g.: Bellis and Gaither 1985; Krause-Jensen et al.,

2004; Ondiviela et al., 2014) the link of these variations with hydrodynamics is still unstudied.

Long-term variations can also be very important. Sea level rise and changes in storms intensity due to climate change represent hazards in the long term. These changes will lead to different flow-ecosystem interactions, and they will change abiotic ecosystem conditions. Then, ecosystems adaptation to climate change and their resilience to extreme events action will be determinant to keep their protection service. In this sense, it is very important to consider the possible hydrodynamic scenarios at the same time ecosystems adaptation is analyzed, e.g. by capturing sediment and building up soil at the same peace water level increases (Leonardi et al., 2015) always considering the potential sediment supply in the area (Fagherazzi et al., 2013). This type of analysis will allow identifying most vulnerable areas to plan future actions to ensure the desired protection. Although some studies have analyzed the potential ecosystems adaptation to climate change effects (e.g.: Boorman, 1992; McCarthy et al., 2011; Gilman et al., 2008) these aspects are still not properly included when evaluating coastal protection services in the long term.

To quantify the flooding risk reduction provided by natural ecosystems under these different scenarios, formulas and parameterizations obtained at smaller scales should feed this larger scale. At this larger scale, laboratory studies are limited due to the impossibility of fiting a complete ecosystem in the experimental facilities. Instead, field campaigns are carried out. They provide with a quantification of wave height attenuation along the ecosystem or flow velocities at some points by using devices, such as pressure sensors and Acoustic Doppler Current Profilers (ADCPs) (e.g.: Mazda et al., 1997; Yang, 1998; Morgan et al., 2009; Glass et al., 2017; Garzon et al., 2019). However, more effort should be done in getting general parameterizations instead of case specific formulas or coefficients. In addition, these parameterizations should be given in terms of ecosystems characteristics linking them with their effect on the hydrodynamic conditions including both, waves and currents.

Large scale numerical models, based on spectral energy conservation or 1-D equations, usually quantify energy attenuation by relying on a drag coefficient or an enhanced bottom friction coefficient (e.g.: Yanagisawa et al., 2009; Zhang et al., 2012; Horstman et al., 2013; Losada et al., 2017).

In most of the cases, C_D is obtained considering formulas relating it to Re or KC. However, as discussed in previous section, these formulas are case dependent and do not include ecosystem properties, such us complex morphology or flexibility. An example is Mendez and Losada (2004) formulation, that was obtained from laboratory experiments performed using seagrass mimics and that has been applied to completely different ecosystems in many studies. Therefore, there is a need to perform a case-by-case calibration of C_D to avoid significant errors in estimating the flood risk reduction provided by the ecosystem (Hortsman et al., 2018). This means that these models are not predictable tools. New approaches based on SVF and HSB that do not rely on the definition of calibration coefficients, as discussed in the previous section, should be further explored to result in predictable numerical tools that can be applied to estimate the energy attenuation provided by ecosystems and ultimately quantify their coastal protection service.

Sediment transport processes are the other key aspects when analyzing the provided coastal protection service. Different field campaigns have been performed using aerial photographs (e.g.: Leonardi et al., 2016) or airborne laser altimetry (e.g.: Temmerman et al., 2005) to quantify sediment elevation. Then, field data have been linked to numerical models. These numerical models parameterize ecosystem effects on sediment transport by means of increasing the erosion threshold reducing the sediment transport in the vegetated area (e.g.: van Maanen et al., 2015; Mariotti and Fagherazzi, 2010). However, a good characterization of these values depending on the type of ecosystem and the forcing hydrodynamic conditions is still needed.

Another important aspect is the potential response of some ecosystems to climate change effects. Some efforts have been done to analyze this potential ecosystem adaptation or loss (e.g.: Mariotti at al., 2010; Fagherazzi et al., 2012; Saintilan et al., 2020). However, these studies are still too theoretical and more practical parameterizations and estimations are needed to be able to include this long-term ecosystem response in the evaluation of protection services for future scenarios.

Considering all the needs described in this section to properly estimate the coastal protection services provided by vegetated ecosystems, Figure 7 shows a summary of the principal gaps identified at the large scale. The main ecological and hydrodynamic parameters involved at this scale and the approaches used to quantify coastal protection services are also presented.

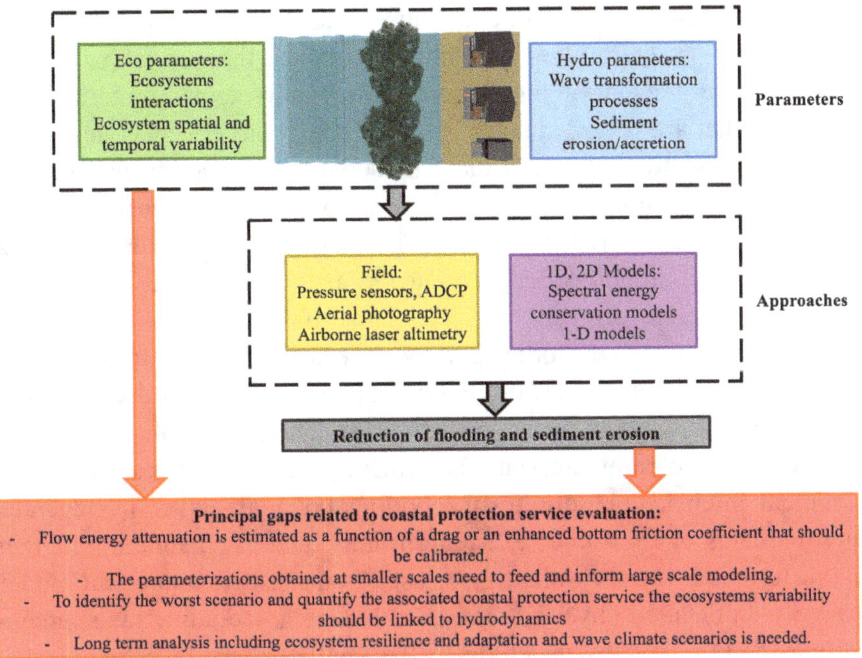

Figure 7. Framework for large scale studies: ecological (green) and hydrodynamics (blue) parameters and different approaches (field campaigns in yellow and numerical models in purple) including main techniques and equations. Outputs obtained from these approaches are presented in grey and principal gaps to be covered in red.

5. Integrating different scales and future needs

As reviewed in previous sections, a huge effort has been made to study flow and sediment interaction with vegetated ecosystems. Many experimental, field and numerical studies related to the estimation of coastal protection services provided by these ecosystems have been carried out at different scales. However, we lack an integrated approach that allows considering all scales and goes from the smallest to the largest scale to first understand the physical processes and then parameterize them to feed into larger scales. The consideration of this integrated approach can benefit from the knowledge gained at different scales to lead to a more accurate and reliable quantification of coastal protection services provided by coastal ecosystems based on the capacity of these ecosystems to reduce flow energy and retain sediment. Thus, results from small scale studies can be used, firstly, to identify all those processes that are relevant to the flow-sediment-vegetation interaction and, secondly, to parameterize those processes that are not adequately resolved at the larger scales, leading to new equations and parameterizations to feed into those larger scale studies.

As described along this chapter, coastal protection services are linked to flood and erosion risk reduction. Therefore, processes related to flow energy attenuation and sediment transport at each scale are of interest to quantify the coastal protection service. Thus, at the small scale, where one or a few individuals are studied and the time scale ranges from seconds to minutes, turbulence processes, bottom boundary layer and drag forces exerted on individuals are of great interest. The results obtained at this scale can lead to parameterizations of these processes, e.g. estimating the turbulent intensity or defining the bottom boundary layer in vegetated environments or the forces exerted on individuals as a function of their characteristics. These results can be used in mesoscale numerical models where such detailed physics are not resolved, e.g. including the source of turbulence produced by vegetation, a bottom boundary layer parameterization for vegetated areas or a momentum sink due to the drag force. In addition, at this community scale, where a community and time scales of minutes to hours are considered, experimental studies provide the estimation of flow energy attenuation along the community and the characterization of sediment transport patterns around and within it. These results are, in turn, necessary inputs to the ecosystem/several ecosystems scale models. This larger scale models, which deal with domains of several kilometers and longtime scales, are used to estimate flood risk reduction based on flow energy attenuation rates and erosion risk reduction based on sediment transport rates. Both rates depend on calibration coefficients, such as the drag coefficient or the bottom friction coefficient that should be characterized in the community scale studies. It is then crucial to move from the small scale to the large scale by understanding and parameterizing the processes necessary to properly define these coefficients and ultimately quantify the coastal protection services provided by NBS including vegetated ecosystems. Figure 8 shows a sketch for an integrated approach considering all different scales and how results from smaller scale studies can feed into larger scale ones. The different spatial and temporal scales are shown in the central panels and the results and parameterizations derived from each scale are shown in the left and right rectangles.

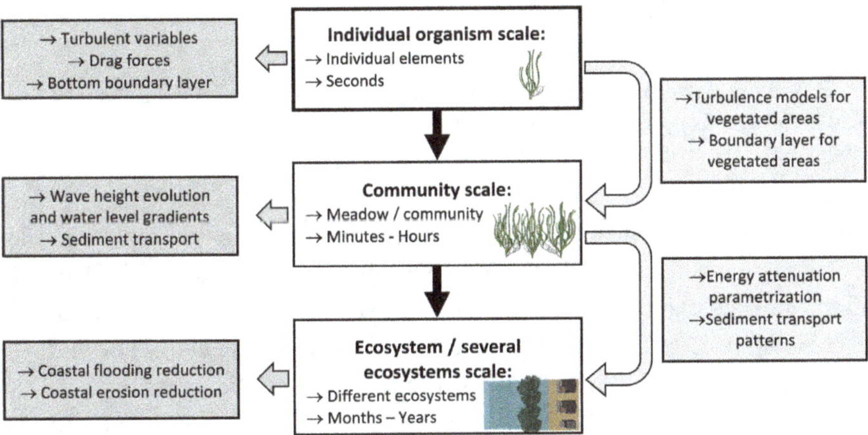

Figure 8. Integrated approach considering different scales: parameterizations needed from each scale to feed the next are displayed in the grey rectangles on the right; results obtained at each scale are shown in the grey rectangles on the left.

Finally, it should be noted that, although an enormous effort has been made in the study of flow-sediment-vegetation interactions in the framework of NBS for coastal protection, as shown in the previous sections, there are still some important needs to adequately quantify coastal protection services. The main gaps identified in the three approaches used to analyze flow-sediment-vegetation interactions and reviewed in this Chapter, are shown in Figure 9. Physical modelling of vegetation ecosystems should include a realistic representation of the ecosystem and the sediment and hydrodynamic conditions by applying appropriate scaling laws if necessary, ensuring similarity between mimics and real individuals if mimics are used, and testing realistic configurations or real vegetation if possible. In the case of field campaigns, it is crucial that flow, sediment

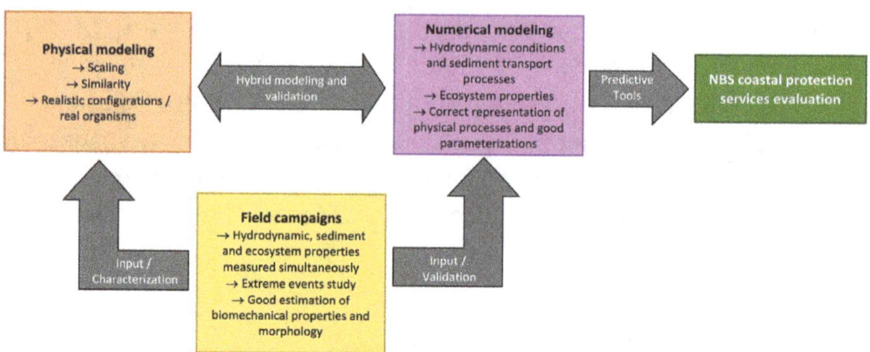

Figure 9. Principal gaps for each approach (physical modeling, numerical modeling and field campaigns) and links between them to ultimately lead to predictive tools for the estimation of NBS coastal protection services.

and ecosystem properties are measured simultaneously, and that biomechanical properties and plant morphology are characterized in detail. In addition, field campaigns can be the main source of data for the estimation of coastal protection services in case of extreme events. The study of extreme events in the laboratory is often limited by the capabilities of the facilities to reproduce extreme events and their dimensions. Therefore, a special effort must be made in the field to characterize the interaction of extreme events with vegetation fields. The main gaps found in the numerical approaches are linked to the adequate reproduction of ecosystem, sediment and flow characteristics and their interactions and to the correct representation of the physical processes that are simulated by using parameterizations. Therefore, special attention must be paid to the correct selection of the numerical tool to be used according to the equations resolved by the model and according to the processes to be analyzed.

These three approaches used to study flow-sediment-vegetation interactions can be complementary. Flow, sediment and ecosystem properties are characterized by field campaigns and these data serve as input to laboratory and numerical analyses. In addition, field and laboratory data are used to validate numerical models. At the same time, numerical models can be used in combination with physical modelling by considering what is known as hybrid modelling, i.e. the combined use of both approaches to study a specific problem. Therefore, the positive feedbacks that can result from considering several approaches can lead to a better understanding and representation of the physical processes involved in the flow-sediment-ecosystem interaction. Figure 9 shows how the different approaches are linked and how they can be combined with each other.

The use of the different approaches based on the ecosystem, sedimentary and hydrodynamic conditions of the study area will ultimately lead to predictive tools for the assessment of NBS coastal protection services and their application. However, further research is still needed to better understand, parameterise and simulate the physical processes to obtain predictive tools involving small uncertainties. The development of these tools will also allow the analysis of temporal and spatial variations in protection services taking into account changes in ecosystem properties, sediment elevation and hydrodynamic conditions. However, long-term and large-scale studies increase uncertainties, which grow with temporal and spatial scale. The current state of knowledge does not yet allow these uncertainties to be limited, and further studies are needed to narrow them down and quantify the protective services provided by NBS based on vegetated coastal ecosystem with a high confidence levels.

References

1. Abdolahpour, M., Ghisalberti, M., McMahon, K. and Lavery, P.S. (2018). *The impact of flexibility on flow, turbulence, and vertical mixing in coastal canopies.* Limnol Oceanogr, 63, 2777–2792.
2. Anderson, M.E., Smith, J.M. (2014). *Wave attenuation by flexible, idealized salt marsh vegetation.* Coastal Engineering, 83, 82–92.
3. Arbele, J., Järvelä, J. (2013). *Flow resistance of emergent rigid and flexible floodplain vegetation.* J. Hydraul. Res. 51(1), 33–45.
4. Augustin, L.N., Irish, J.L., Lynett, P. (2009). *Laboratory and numerical studies of wave damping by emergent and near emergent wetland vegetation.* Coastal Engineering 56, 332–340.
5. Bellis, V.J., Gaither, A.C. (1985). *Seasonality of aboveground and belowground biomass for six salt marsh plant species.* The Journal of the Elisha Mitchell Scientific Society, 101(2), 95–109.
6. Boorman, L.A. (1992). *The environmental consequences of climatic change on British saltmarsh vegetation.* Wetlands Ecology and Management, 2, 11–21.

7. Bouma, T.J., Friedrichs, M., Van Wesenbeeck, B.K., Temmerman, S., Graf, G., Herman, P.M.J. (2009). *Density-dependent linkage of scale-dependent feedbacks: A flume study on the intertidal macrophyte Spartina anglica.* Oikos, 118: 260–268.

8. Bouma, T.J., De Vries, M.B., Herman, P.M.J. (2010). *Comparing ecosystem engineering efficiency of two plant species with contrasting growth strategies.* Ecology 91(9), 2696–2704.

9. Bouma, T.J., Temmerman, S., van Duren, L.A., Martini, E., Vandenbruwaene, W., Callaghan, D.P., Balke, T., Biermans, G., Klaassen, P.C., van Steeg, P., Dekker, F., van de Koppel, J., de Vries, M.B., Herman, P.M.J. (2013). *Organism traits determine the strength of scale-dependent bio-geomorphic feedbacks: A flume study on three intertidal plant species.* Geomorphology 180–181, 57–65.

10. Bouma, T.J., van Belzen, J., Balke, T., Zhu, Z., Airoldi, L., Blight, A.J., Davies, A.J., Galvan, C., Hawkins, S.J., Hoggart, S.P.G., Lara, J.L., Losada, I.J., Maza, M., Ondiviela, B., Skov, M.W., Strain, E.M., Thompson, R.C., Yang, S., Zanutigh, B., Zhang, L. and Herman, P.M.J., (2014). *Identifying knowledge gaps hampering application of intertidal habitats in coastal protection: Opportunities & steps to take.* Coastal Engineering, 87: 147–157

11. Callaghan, F., Cooper, G., Nikora, V., Lamouroux, N., Statzner, B., Sagnes, P., Radford, J., Malet, E., Biggs, B. (2007). *A submersible device for measuring drag forces on aquatic plants and other organisms.* New Zeal J Mar Freshwat Res, 41, 119–127

12. Chapman, J.A., Wilson, B.N., Gulliver, J.S. (2015). *Drag force parameters of rigid and flexible vegetal elements.* Water Resources Research, 51, 3292–3302.

13. Chen, S.-N., Sanford, L.P., Koch, E.W., Shi, F., North, E.W. (2007). *A nearshore model to investigate the effects of seagrass bed geometry on wave attenuation and suspended sediment transport.* Estuaries and Coasts, 30, 296–310.

14. Christianen, M.J.A., van Belzen, J., Herman, P.M.J., van Katwijk, M.M., Lamers, L.P.M., van Leent, P.J.M., Bouma, T.J. (2013) *Low-canopy seagrass beds still provide important coastal protection services.* PLoS ONE, 8(5), e62413.

15. Cui, J., Neary, V.S. (2002). *Large eddy simulation (LES) of fully developed flow through vegetation.* Hydroinformatics 2002: Proc. Fifth Int. Conf. Hydroinformatics, Cardiff, U. K., pp. 39–44.

16. Duarte, C.M., Losada, I.J., Hendriks, I., Mazarrasa, I., Marba, N. (2013). *The potential of coastal habitats for climate change mitigation and adaptation.* Nat. Clim. Chang. 3, 961–968.

17. Fagherazzi S., Kirwan, M.L., Mudd, S.M., Guntenspergen, G.R., Temmerman, S., D'Alpaos, A., van de Koppel, J., Rybczyk, J.M., Reyes, E., Craft, C., Clough, J., (2012). *Numerical models of salt marsh evolution: Ecological and climatic factors.* Reviews of Geophysics, 50, 1.

18. Fagherazzi, S., Mariotti, G., Wiberg, P.L., McGlathery, K.J. (2013). *Marsh collapse does not require sea level rise.* Oceanography, 26(3), 70–77.

19. Ferrario F., Beck M., Storlazzi C., Micheli F., Shepard C., and Airoldi L. (2012). *The effectiveness of coral reefs for coastal hazard risk reduction and adaptation.* Nature Communications 5. Article number 3789.

20. Frostick, L.E., Thomas, R.E., Johnson, M.F., Rice, S.P., McLelland, S.J. (2014). *Users Guide to Ecohydraulic Modelling and Experimentation: Experience of the Ecohydraulic Research Team (PISCES) of the HYDRALAB Network.* IAHR Design Manual.

21. Garzon, J.L., Maza, M., Ferreira, C.M., Lara, J.L., Losada, I.J. (2019). *Wave attenuation by Spartina saltmarshes in the Chesapeake Bay under storm surge conditions.* Journal of Geophysical Research: Oceans, 124(7), 5220–5243.

22. Gaylord, B., Denny, M.W. (2003). *Modulation of wave forces on kelp canopies by along-shore currents.* Limnol. Oceanogr. 48, 860–871.

23. Ghisalberti, M., Nepf, H.M. (2002). *Mixing layers and coherent structures in vegetated aquatic flows.* J. Geophys. Res. Oceans, 107(C2), 3-1-3-11.

24. Gillis, L.C, Maza, M., Garcia-Maribona, J., Lara, J.L., Suzuki, T., Arguemi, M., Paul, M., Folkard, A., Balke, T. (2022) *Living on the edge: How traits of ecosystem engineers drive bio-physical interactions at coastal wetland edges.* Advances in Water Resources, 166, 104257.

25. Gilman, E.L., Ellison, J., Duke, N.C., Field, C. (2008). *Threats to mangroves from climate change and adaptation options: A review.* Aquatic Botany, 89(2), 237–250.

26. Gosselin, F., De Langre, E., Machado-Almeida, B.A. (2010). *Drag reduction of flexible plates by reconfiguration.* J. Fluid Mech., 650, 319–341.

27. Hiraoka, H., Ohashi, M., 2006. *A (k–ε) turbulence closure model for plant canopy flows.* Proc. of the 4th International Symposium on Computational Wind Eng. (CWE2006) in Yokohama.

28. Horstman, E.M., Dohmen-Janssen, C.M., Hulscher, S.J.M.H., 2013. *Modeling tidal dynamics in a mangrove creek catchment in Delft3D.* Coastal Dynamics 2013 Proceedings 833–844.

29. Horstman, E.M., Dohmen-Janssen, C.M., Narra, P.M.F., van den Berg, N.J.F., Siemerink, M., Hulscher, S.J.M.H., 2014. *Wave attenuation in mangroves: A quantitative approach to field observations.* Coast. Eng. 94, 47–62.

30. Horstman, E.M., Bryan, K.M., Mullarney, J.C., Pilditch, C.A., Eager, C.A., 2018. *Are flow-vegetation interactions well represented by mimics? A case study of mangrove pneumatophores.* Adv. Water Resour. 111, 360–371.

31. Hu, Z., Suzuki, T., Zitman, T., Uittewaal, W., Stive, M., 2014. *Laboratory study on wave dissipation by vegetation in combined current–wave flow.* Coast. Eng. 88, 131–142.

32. Jalonen, J., Järvelä, J., Aberle, J. (2013). *Leaf area index as vegetation density measure for hydraulic analyses.* J. Hydraul. Eng., 139(5), 461–469.

33. Kim, S. J., Stoesser, T. (2011). *Closure modeling and direct simulation of vegetation drag in flow through emergent vegetation.* Water Resour. Res., 47, W10511.

34. King, A.T., Tinoco, R.O., Cowen, E.A. (2012). *A k-ε turbulence model based on the scales of vertical shear and stem wakes valid for emergent and submerged vegetated flows*. Journal of Fluid Mechanics, 701, 1–39.

35. Krause-Jensen, D., Diaz, E., Cunha, A.H., Greve, T.M. (2004). *Have seagrasses distribution and abundance changed? European seagrasses: An introduction to monitoring and management*. The M&MS project. Monitoring and Managing of Europan Seagrasses EVK3-CT-2000-00044.

36. Lara, J.L., Losada, I.J., Maza, M., Jaime, F.J. (2015). *Hybrid/Composite modelling: Combined physical-numerical modelling*. 7th International Short Course and Conference on applied coastal research, 28 Sept.-1 Oct. 2015, Florence, Italy.

37. Lara, J.L., Maza, M., Ondiviela, B., Trinogga, J., Losada, I.J., Bouma, T.J., Gordejuela, N. (2016). *Large-scale 3-D experiments of wave and current interaction with real vegetation. Part 1: Guidelines for physical modelling*. Coastal Engineering, 107, 70–83.

38. Larsen, L.G., Harvey, J.W. (2010). *How vegetation and sediment transport feedbacks drive landscape change in the everglades and wetlands worldwide*. The American Naturalist, 176(3), E66-79.

39. Leonardi, N., Ganju N., K., Fagherazzi, S. (2015*). A linear relationship between wave power and erosion determines salt-marsh resilience to violent storms and hurricanes*. Proceeding of the National Academy of Science.

40. Li, C.W., Yu, L.H. (2010). *Hybrid LES/RANS modelling of free surface flow through vegetation*. Computers and Fluids, 39(9), 1722–1732.

41. Li, C.W., Xie, J.F. (2011). *Numerical modeling of free surface flow over submerged and highly flexible vegetation*. Advances in Water Resources, 34(4), 468–477.

42. Liu, D., Diplas, P., Fairbanks, J., Hodges, C. (2008). *An experimental study of flow through rigid vegetation*. J. Geophys. Res., 113, F04015.

43. Lopez, F., Garcia, M. (2001). *Mean flow and turbulence structure of open-channel flow through non-emergent vegetation*. J. Hydraul. Eng., 127(5), 392–402.

44. Losada, I.J., Menéndez, P., Espejo, A., Torres-Ortega, S., Díaz-Simal, P., Fernández, F., Abad, S., Ripoll, N., García, J., Beck, M.W., Narayan, S., Trespalacios, D.M., Quiroz, A., 2017. *Valuing protective services of mangroves in the Philippines: Technical report*. World Bank Group, Washington, D.C.

45. Luhar, M., Coutu, S., Infantes, E., Fox, S., Nepf, H. (2010). *Wave-induced velocities inside a model seagrass bed*. Journal of Geophysical Research, 115, C12005.

46. Luhar, M., Nepf, H. (2011). *Flow-induced reconfiguration of buoyant and flexible aquatic vegetation*. Limnol. Oceanogr., 56(6), 2003–2017.

47. McCarthy, J.J., Canziani, O., Leary, N.A., Dokken, D.J. and White, K.S., eds (2001) *Climate Change 2001: Impacts, Adaptation, and Vulnerability*. New York: Cambridge University Press, 1032.

48. Mariotti, G., Fagherazzi, S. (2010). *A numerical model for the coupled long-term evolution of salt marshes and tidal flats*. J. Geophys. Res., 115, F01004.

49. Mariotti, G., Fagherazzi, S., Wiberg, P.L., McGlathery, K.J., Carniello, L., Defina, A. (2010). *Influence of storm surges and sea level on shallow tidal basin erosive processes.* J. Geophys. Res., 115, C11012.

50. Marois, D.E., Mitsch, W.J. (2015). *Coastal protection from tsunamis and cyclones provided by mangrove wetlands – A review.* International Journal of Biodiversity Science, Ecosystem Services & Management, 11:1, 71-83.

51. Maza, M., Lara, J.L., Losada, I.J. (2013a). *A coupled model of submerged vegetation under oscillatory flow using Navier–Stokes equations.* Coastal Engineering, 80, 16–34.

52. Maza, M., Lara, J.L., Losada, I.J. (2013b). *Flow patterns around vegetation meadows.* Proceedings of the 8th River, Coastal and Estuarine Morphodynamics (RCEM), Santander, Spain.

53. Maza, M., Lara, J.L., Losada, I.J., Ondiviela, B., Trinogga, I.J., Bouma, T.J. (2015a) *Large-scale 3-D experiments of wave and current interaction with real vegetation. Part 2: Experimental analysis.* Coastal Engineering, 106, 73–86.

54. Maza, M., Lara, J.L., Losada, I.J., (2015b). *Tsunami wave interaction with mangrove forests: a 3-D numerical approach.* Coastal Engineering, 98, 33–54.

55. Maza, M., Lara, J.L., Losada, I.J. (2016) *Solitary wave attenuation by vegetation patches.* Advances in Water Resources, 98, 159–172.

56. Maza, M., Lara, J.L., Losada, I.J (2019) *Experimental analysis of wave attenuation and drag forces in a realistic fringe Rhizophora mangrove forest.* Advances in Water Resources, 131, 103376.

57. Maza, M., Lara, J.L., Losada, I.J. (2022) *A paradigm shift in the quantification of wave energy attenuation due to saltmarshes based on their standing biomass.* Scientific Reports, 12(1), 13883.

58. Mazda, Y., Magi, M., Kogo, M., Hong, P.N. (1997). *Mangroves as a coastal protection from waves in the Tong King delta, Vietnam.* Mangroves and Salt Marshes, 1, 127–135.

59. McIvor, A., Möller, I., Spencer, T., Spalding, M. (2012a). *Reduction of wind and swell waves by mangroves.* Natural Coastal Protection Series: Report 1. The Nature Conservancy and Wetlands International.

60. McIvor, A., Spencer, T., Möller, I., Spalding, M. (2012b). *Storm surge reduction by mangroves.* Natural Coastal Protection Series: Report 2. The Nature Conservancy and Wetlands International.

61. Möller, I., Kudella, M., Rupprecht, F., Spencer, T., Paul, M., van Wesenbeeck, B., Wolters, G., Jensen, K., Bouma, T.J., Miranda-Lange, M., Schimmels, S. (2014). *Wave attenuation over coastal salt marshes under storm surge conditions.* Nature Geoscience, 7.

62. Montakhab, A., Yusuf, B., Ghazali, A., Mohamed, T. (2012). *Flow and sediment transport in vegetated waterways: A review.* Rev. Environ. Sci. Bio/Technol., 11(3), 275–287.

63. Morgan, P.A., Burdick, D.M., Short, F.T. (2009). *The functions and values of fringing salt marshes in northern New England, USA.* Estuaries and Coasts 32, 483–495.

64. Morris, J.T., Sundareshwar, P., Nietch, C.T., Kjerfve, B. and Cahoon, D. (2002). *Responses of coastal wetlands to rising sea level.* Ecology 83, 2869–2877.

65. Narayan, S., Beck, M.W., Reguero, B.G., Losada, I.J., van Wesenbeeck, B., Pontee, Sanchirico, J.N., Ingram, J.C., Lange, G.M., Burks-Copes, K.A. (2016). *The effectiveness, costs and coastal protection benefits of natural and nature-based defences.* PLoS ONE, 11(5), e0154735.

66. Nardin, W., Edmonds, D.A. (2014). *Optimum vegetation height and density for inorganic sedimentation in deltaic marshes.* Nature Geoscience, 7, 722–726.

67. Nepf, H.M. (1999). *Drag, turbulence and diffusion in flow through emergent vegetation.* Water Resour. Res., 35, 479–489.

68. Nepf, H.M. (2012). *Hydrodynamics of vegetated channels.* Journal of Hydraulic Research, 50:3, 262–279.

69. Nestler, J.M., Stewardson, M.J., Gilvear, D.J., Webb, A., Smith, D.L. (2016). *Ecohydraulics exemplifies the emerging "paradigm of the interdisciplines".* Journal of Ecohydraulics, 1:1-2, 5–15.

70. Nezu, I., Nakagawa, H. (1993). *Turbulence in Open-Channel Flows.* Balkema, Rotterdam, Netherlands.

71. Okamoto, T., Nezu, I., Sanjou, M. (2016). *Flow–vegetation interactions: Length-scale of the "monami" phenomenon.* Journal of Hydraulic Research, 54(3), 251–162.

72. Ondiviela, B., Losada, Lara, J.L., I.J., Maza, M., Galvan, C., Bouma, T., van Belzen, J. (2014). *The role of seagrasses on coastal protection in a changing climate.* Coastal Engineering. 87, 158–168.

73. Ozeren, Y., Wren, D.G., Wu, W. (2013). *Experimental investigation of wave attenuation through model and live vegetation.* J. Waterway, Port, Coastal, Ocean Eng., 04014019.

74. Palau-Salvador, G., Stoesser, T., Rodi, W. (2008). *LES of the flow around two cylinders in tandem.* Journal of Fluids and Structures 24(8):1304–1312.

75. Paul, M., Bouma, T.J., Amos, C.L. (2012). *Wave attenuation by submerged vegetation: Combining the effect of organism traits and tidal current.* Mar. Ecol. Prog. Ser. 444, 31–41.

76. Peralta, G., Brun, F.G., Perez-Llorens, J.L., Bouma, T.J. (2006). *Direct effects of current velocity on the growth, morphometry and architecture of seagrasses: A case study on Zostera noltii.* Mar. Ecol. -Prog. Ser. 327, 135–142.

77. Puijalon, S., Bornette, G., Sagnes, P. (2005). *Adaptations to increasing hydraulic stress: Morphology, hydrodynamics and fitness of two higher aquatic plant species.* J. Exp. Bot. 56, 777–786.

78. Saintilan, N., Khan, N.S., Ashe, E., Kelleway, J.J., Rogers, K., Woodroffe, C.D., Horton, B. P., (2020). *Thresholds of mangrove survival under rapid sea level rise.* Science 118–1121.

79. Shepard, C.C., Crain, C.M., Beck, M.W., 2011. *The protective role of coastal marshes: A systematic review and meta-analysis.* PLoS One 6, e27374.

80. Shucksmith, J.D., Boxall, J.B., Guymer, I. (2010). *Effects of emergent and submerged natural vegetation on longitudinal mixing in open channel flow.* Water Resour. Res., 46, W04504.

81. Singh, R., Bandu, M.M., Mahadevan, A., Mandre, S. (2015). *Monami as an oscillatory hydrodynamic instability in a submerged sea grass bed.* Journal of Fluid Mechanics, 786.

82. Sonnenwald, F., Hart, J.R., West, P., Stovin, V.R., Guymer, I. (2017). *Transverse and longitudinal mixing in real emergent vegetation at low velocities.* Water Resour. Res., 53.

83. Stewart, H. L. (2006). *Morphological variation and phenotypic plasticity of buoyancy in the macroalga Turbinaria ornata across a barrier reef.* Mar. Biol. 149, 721–730.

84. Stoesser, T., Kim, S.J., Diplas, P. (2010). *Turbulent flow through idealized emergent vegetation.* J. Hydraul. Eng., 136(12), 1003–1017.

85. Stratigaki, V., Manca, E., Prinos, P., Losada, I.J., Lara, J.L., Sclavo, M., Amos, C.L., Cáceres, I., Sánchez-Arcilla, A. (2011). *Large-scale experiments on wave propagation over Posidonia oceanica.* J. of Hydraulic Res., 49, 31–43.

86. Strusinska-Correia, A., Husrin, S., Oumeraci H. (2013). *Tsunami damping by mangrove forest: A laboratory study using parameterized trees.* Nat. Hazards Earth Syst. Sci., 13, 483–503.

87. Suzuki, T., Zijlema, M., Burger, B., Meijer, M.C., Narayan, S. (2012). *Wave dissipation by vegetation with layer schematization in SWAN.* Coast. Eng. 59 (1), 64–71.

88. Tanaka, N., Sasaki, Y., Mowjood, M.I.M., Jinadasa, K.B.S.N., Homchuen, S. (2007). *Coastal vegetation structures and their functions in tsunami protection: Experience of the recent Indian Ocean tsunami.* Landsc. Ecol. Eng. 3, 33–45.

89. Temmerman, S., T.J. Bouma, G. Govers, Z.B. Wang, M.B. De Vries, and P.M.J. Herman. (2005). *Impact of vegetation on flow routing and sedimentation patterns: Three-dimensional modeling for a tidal marsh.* J. Geophys. Res., 110, F04019.

90. Temmerman, S., Meire, P., Bouma, T.J., Herman, P.M.J., Ysebaert, T., De Vriend, H.J. (2013). *Ecosystem-based coastal defence in the face of global change.* Nature 493, 45–49.

91. Tinoco, R.O., Cowen, E.A. (2013). *The direct and indirect measurement of boundary stress and drag on individual and complex arrays of elements.* Exp Fluids, 54:1509, 1–16.

92. Tonelli, M., Fagherazzi, S., Petti, M. (2010). *Modeling wave impact on salt marsh boundaries.* J. Geophys. Res., 115, C09028.

93. van Maanen, B., Coco, G., Bryan, K.R. (2015). *On the ecogeomorphological feedbacks that control tidal channel network evolution in a sandy mangrove setting.* Proc. Math Phys Eng Sci., 471(2180), 20150115.

94. van Rooijen, A.A., McCall, R.T., van Thiel de Vries, J.S.M., van Dongeren, A.R., Reniers, A.J.H.M., and Roelvink, J.A. (2016). *Modeling the effect of wave-vegetation interaction on wave setup.* J. Geophys. Res. Oceans, 121, 4341–4359

95. Widdows, J., Pope, N.D., Brinsley, M.D. (2008). *Effect of Spartina Anglica Stems on near-Bed Hydrodynamics, Sediment Erodability and Morphological Changes on an Intertidal Mudflat.* Marine Ecology Progress Series 362: 45–57.

96. Wu, W., Shields, D. Jr., Bennett, S.J., Wang, S.S.Y. (2005). *A depth-averaged two-dimensional model for flow, sediment transport, and bed topography in curved channels with riparian vegetation.* Water Res. Res., 41, W03015.

97. Yanagisawa, H., Koshimura, S., Goto, K., Miyagi, T., Imamura, F., Ruangrassamee, A., Tanavud, C. (2009). *The reduction effects of mangrove forest on a tsunami based on field surveys at Pakarang Cape, Thailand and numerical analysis.* Estuarine, Coastal and Shelf Science 81, 27–37.

98. Yang, J.Q., Kerger, F., Nepf, H. (2015). *Estimation of the bed shear stress in vegetated and bare channels with smooth beds.* Water Resour. Res., 51, 3647–3663.

99. Zhang, K., Liu, H., Li, Y., Xu, H., Shen, J., Rhome, J., Smith III, T.J. (2012). *The role of mangroves in attenuating storm surges.* Estuarine, Coastal and Shelf Science 102, 11–23.

100. Zhou, Z., Coco, G., Jimenez, M., Olabarrieta, M., van der Wegen, M., Townend, I. (2014). *Morphodynamics of river-influenced back-barrier tidal basins: The role of landscape and hydrodynamic settings.* Water Resources Research, 50(12), 9514–9535.

101. Zong, L., Nepf, H. (2010). *Flow and deposition in and around a finite patch of vegetation.* Geomorphology, 116(3), 363–372.

Chapter 5

Installation of a Hybrid Tsunami Defense System as a Combination of Engineered Structures and Coastal Forest and Vegetation Management

Norio Tanaka[1,*], Yoshiya Igarashi[1] and Hiroyuki Torita[2]

[1]*Graduate School of Science and Engineering, Saitama University,
255 Shimo-okubo, Sakura-ku, Saitama 338-8570, Japan*
**tanaka01@mail.saitama-u.ac.jp*

[2]*Hokkaido Research Organization,
Koshunai, Bibai, Hokkaido 079-0198, Japan*

The advantage (reduction of fluid forces) and disadvantage (driftwood production) of a coastal forest as a tsunami buffer zone are summarized. Considering the functional limitations and disadvantages, a hybrid defense system (HDS) that strengthens existing coastal forests as a mitigation measure for a future large tsunami by combining a forest with an embankment and/or a moat has been proposed. Utilizing knowledge obtained from flume experiments and numerical simulations, two hybrid defense systems (HDSs), VME (vegetation (V), moat (M), and embankment (E) from seaward) and EMV, the reverse order of VME, have been installed in Japan based on the existing forest conditions and land availability. For the coastal forest construction, stand structures of vegetation (density, diameter, tree crown height, etc.) and their critical breaking values are important for increasing the mitigation effect. Forest management (in forestry, thinning) should be discussed based on the parameters in each species.

1. Introduction

The Indian Ocean Tsunami (IOT) in 2004 and the Great East Japan tsunami (GEJT) in 2011 caused tremendous damage to coastal communities and infrastructures, and had an enormous economic and environmental impact in the Asian region. Since the IOT, an early warning system and plans for evacuation from high-risk areas have been discussed as necessary countermeasures for disaster mitigation of tsunamis. In addition, many structural methods for disaster mitigation have been proposed. They can be categorized as non-structural measures (i.e., vegetation buffer zones, sand dunes, lagoons, coral reefs, and/or combinations) and structural measures (i.e., sea walls, large embankments, gates, and evacuation buildings). Especially since the Papua New Guinea tsunami in 1998 and the 2004 IOT, many studies have identified the effectiveness (i.e., fluid force reduction, secondary damage mitigation by trapping debris) and the disadvantages caused by the functional limitations (destruction of trees, production of driftwood, as the cause of secondary damage) of a coastal forest as a barrier to mitigate tsunami destruction (ex., Hirashi & Harada, 2003; Danielsen et al., 2005; Kathiresan & Rajendran, 2005; Tanaka et al., 2007; Yanagisawa et al., 2009). The advantages and disadvantages of a coastal forest on tsunami or storm surge impact were investigated by post disaster surveys, water flume experiments, and/or numerical simulations.

Under the restrictions of land availability or the destruction by fluid forces for keeping a wide enough vegetation buffer for large tsunamis, a sea embankment followed by a coastal forest could have a higher mitigation function against an inundating current (Tanaka et al., 2014). After the 2011 GEJT, tsunamis were classified into two categories in Japan, i.e., Level 1 (disaster prevention level) and Level 2 (disaster mitigation level) in which the recurrent period is less than around a hundred years and several hundred to a thousand years, respectively, by the Ministry of Land, Infrastructure, Transport and Tourism, Japan (MLIT). The embankment height is targeted or has already been constructed for a Level 1 tsunami, but disaster mitigation methods need to be developed for a Level 2 tsunami. A combination of natural and artificial measures is recommended to mitigate a tsunami by enhancing coastal resilience in the protection from or mitigation of storm and coastal flooding (Temmerman et al., 2013; Tanaka et al., 2014; Sutton-Grier et al., 2015). After the GEJT, studies of a hybrid solution as a combination of engineered structures and coastal vegetation, a hybrid defense system (HDS), were started that include our research group (Igarashi et al., 2018; Igarashi & Tanaka, 2018; Zaha et al.,

2019; Rashedunnabi and Tanaka, 2020). In particular, Zaha et al. (2019) conducted flume experiments to elucidate the optimal arrangement of the HDS in Hokkaido Prefecture, Japan. The two optimal structures in the study are now installed in a coastal region in Hokkaido Pref. as a pilot project for mitigating future tsunamis.

This chapter reviews and discusses 1) the advantages (reduction of tsunami fluid forces and delay in tsunami arrival) and disadvantages (driftwood production due to the destruction of trees and secondary damages), 2) the effectiveness of HDS as revealed by flume experiments and numerical simulations, 3) introduction of HDS in Hokkaido Pref., and 4) the importance of forest management.

2. Effectiveness and vulnerability of coastal forests against tsunami revealed by post tsunami surveys

A coastal forest was used as a natural buffer zone more than 100 years ago for mitigating disasters in Japan (Honda, 1898). However, the effects of coastal forests were limited under a large destructive tsunami with great water depth, large velocity, and strong bed shear stress (Shuto et al., 1987). After the 1998 Papua New Guinea tsunami, Dengler and Preuss (2003) pointed out the relatively greater resistance of *Casuarina* trees than that of palm trees and the need to conduct further studies to clarify the disadvantages of *Casuarina* trees when they became driftwood by scouring in order to utilize vegetation as a tsunami buffer zone. After the 2004 IOT, the importance of a mangrove forest as a buffer zone was reported by comparing the difference in the disaster conditions between exposed villages and those behind mangrove forests (Danielsen et al., 2005).

2.1. *Importance of stand structure of coastal forest and its management*

Tanaka et al. (2007) showed evidence by a post-tsunami survey of IOT that tree trunk diameter at breast height and tree density could dependently affect the energy reduction and the destructiveness of a tsunami because trees with larger trunk diameters need more space between them for growth. They also elucidated the importance of tree crown height with respect to tsunami height for quantitative understanding of tree resistance and destruction. This suggested the importance of studying the forest structure of different species and forest management to effectively mitigate a tsunami. The United Nations Food and Agriculture Organization (FAO) (2007) also recognized the importance of producing stands of tree of various ages with a wide range of sizes to enhance the mitigation potential because the potential of a forest of a single species declines due to self-thinning and the increase in tree crown height with age.

For increasing the resistance of coastal forest, a combination of multiple tree species that have different stand structures was recommended for a buffer forest (Tanaka et al., 2007). Especially, dense vegetation like *Pandanus odoratissimus* is suitable for providing strong resistance, and large diameter trees like *Casurina equisetifolia* can be provided for trapping tsunami-borne floating debris. Coastal trees also have other roles, i.e., providing something for people to climb or a soft-landing place for washed-out people (Shuto, 1987; Tanaka et al., 2007). Broad-leaved trees were reported effective and should be grown behind this buffer forest (Tanaka et al., 2007). These studies demonstrated the importance of paying more attention to the stand structures (tree density, trunk diameter, trunk height, crown height) of representative tree species for providing resistance, trapping debris, and emergency evacuation. This is also discussed in Section 5 for the practical implementation of hybrid defense system.

Soil resistance is also an important factor for maintaining the resisting effects of coastal forest. Scouring due to poor soil resistance can not only reduce the vegetative resistance but produce secondary damage after the washout of trees (Shuto, 1987; Tanaka et al., 2007; Dengler and Preuss, 2003).

In the 2011 GEJT, much driftwood was produced by a coastal forest just behind a coastal embankment where severe scouring occurred (Tanaka et al., 2013, 2014). When the forest was located on a sand dune, the landward side of the coastal forest on the downslope of sand dunes produced floating driftwood where the shear stress increased because of the change of slope angle (Ali and Tanaka, 2020). Based on that evidence, an inland forest for trapping debris was proposed (Tanaka et al., 2013).

There is a long-running argument about the effectiveness of a coastal forest as a buffer zone because it increases the secondary damage to buildings by impact force or additional drag force when it has accumulated in front of buildings. Tanaka and Onai (2016) demonstrated that the trapping of driftwood by large diameter standing trees could greatly decrease the tsunami fluid force even further inland in the Sendai Plain at the GEJT, and reduce the percentage of washed-out houses from a numerical simulation and the fragility curve in the district. Tanaka and Ogino (2017) compared the two controversial aspects, the advantage of reducing the fluid force and the disadvantage of increasing the impact force or additional resistance when the driftwood accumulated in front of a house. The existence of a coastal forest can reduce the fluid force itself. When a collision between a house and pieces of driftwoods occurs just near the peak of the fluid force of the inundating tsunami current, it has some possibility to increase the total fluid force; however, most of the collisions do not occur at the same time. Further, the additional fluid force due to damming in front of a house was found to be smaller in most of the cases. So, the disadvantage is less when a forest can reduce the tsunami fluid force. The papers proposed to utilize more vegetation for disaster mitigation together with civil engineering structures.

Because large diameter trees that can survive the destructive force of a tsunami (Tanaka et al., 2013) need large spaces for growth, it is sometimes difficult to create a dense forest, and thus, it is not easy to provide a large resistance. Based on the results of Tanaka and Onai (2016), Rahman et al. (2022) proposed planting small trees in the locations where scouring occurs. In that case, when driftwood production occurs downstream of the embankment and trapping occurs at the landward forest, a large energy reduction can be expected. It can also decrease the scour depth at the embankment toe and increase the stability of the embankment itself. This is a new idea for utilizing driftwood for disaster mitigation and should be investigated further using large scale experimental facilities and numerical simulations.

2.2. *Breaking condition of trees*

Post-tsunami surveys have revealed the breaking conditions of trees. Shuto (1987) investigated the records of five previous Japanese tsunamis (in 1896, 1933, 1944, 1960, and 1983) and classified the degrees of forest damage. The forest damage conditions were statistically related to the tsunami inundation depths and the forest characteristics. The study introduced an important parameter for discussing the degree of forest damage with tsunami extent, the vegetation thickness 'dn', where d is the tree trunk diameter at breast height (cm), n is the number of trees in a forest in the streamwise direction and in unit widths in the cross-stream direction. As the parameter dn did not directly include aspects of the tree stand structure like tree height and tree crown height, Tanaka et al. (2007) extended the parameter to include the differences of tree stand structures by integrating the tree shape effect.

There are mainly three breaking mechanisms of a tree: tree trunk breakage, tree overturning, and washout by scouring. Tanaka et al. (2018) included the critical conditions of tree trunk breakage and tree overturning and the change in vegetation resistance, as seen in Fig. 1. The occurrence of tree trunk breakage was modelled as the moment when the fluid force (M_x) acting on a tree trunk at each height exceeds the critical value of the strength of the tree trunk. Tree overturning was also modelled to occur when M_x exceeds the critical overturning moment, which is related to the root weight (Samarakoon et al., 2013), and thus tree weight (Tanaka et al., 2018).

The critical breaking moment of a tree trunk at X (M_{criX}) was usually analyzed as a cubic function of the tree trunk diameter:

$$M_{criX} = \frac{\pi d_X^3}{32}\sigma_{MAX} \tag{1}$$

where σ_{MAX} is the critical stress value for breaking the tree trunk, and d_X is tree trunk diameter at X.

The critical overturning moment of a tree (M_{criOT}) was usually analyzed by a function of tree weight (diameter^2×tree height):

$$M_{criOT} = k_O d_m^2 H_t + k_C \tag{2}$$

where H_t and d_m are tree height and tree trunk diameter at breast height, respectively. k_O and k_C are site- and species-specific coefficients which need to be determined by field tests because they are greatly affected by soil strength and root volume (Samarakoon et al., 2013; Tanaka et al., 2015).

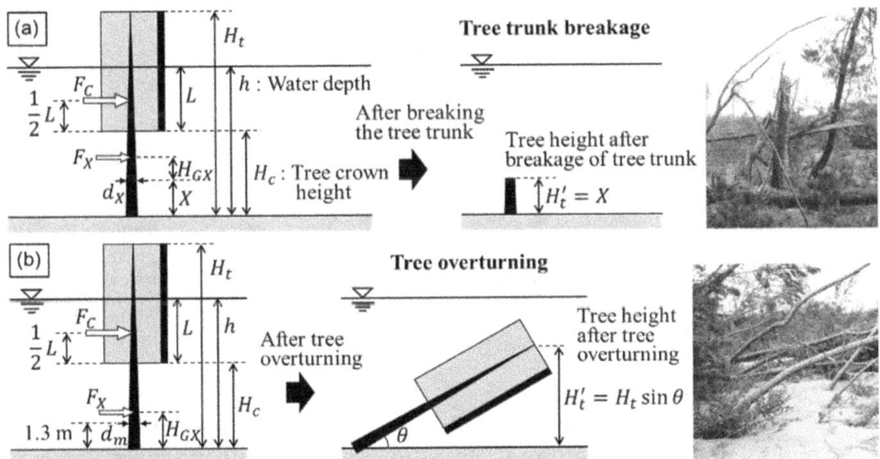

Fig. 1. Representative tree breaking mode included in numerical simulation. (a) Tree-trunk breakage, (b) overturning (modified from Tanaka et al. (2018)).

3. Implications of effectiveness of hybrid structure comprising a coastal forest and embankment in the 2004 GEJT

Tanaka et al. (2014) conducted a post GEJT survey in the Sendai Plain, Japan, to investigate the damage to a coastal forest and wood buildings (houses), and analyzed the advantage of the example of a hybrid structure of an embankment followed by a coastal forest on the reduction of washed out houses by a nonlinear long wave equation model that included the resistance change due to tree breaking. The simulations reproduced the effects of an approximately 10-m tsunami at the coast and the length of the damage to a coastal forest, and demonstrated that the washout length of houses by the coastal forest (600 m in width) and coastal embankment (5.4–6.4 m in height) were reduced by approximately 100 m and 560–1520 m, respectively.

　　　　　　　N. Tanaka, Y. Igarashi and H. Torita

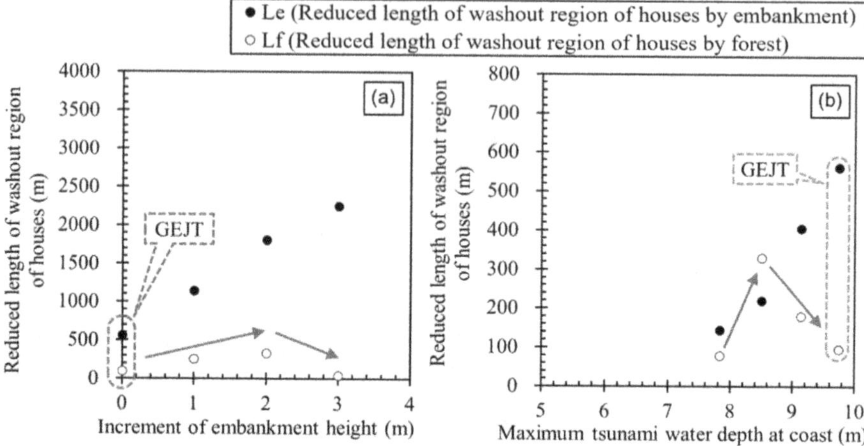

Fig. 2. Reduction of the length of washout region of houses by forest and by embankment with changing. (a) Embankment height, (b) tsunami water depth at coast (modified from Tanaka et al. (2014)).

Because the reduction of damage by the coastal forest was relatively less than that of an embankment, the study further analyzed the mitigation effect by changing embankment height and tsunami water depth at the coast (Fig. 2). With increasing embankment height from the height of the GEJT (Tokyo Peil (T.P.) 5.4 m), the washout length of houses, estimated by a fragility curve for washing out houses as a function of fluid force, was greatly reduced not only by embankment height but also by vegetation resistance. Increasing the embankment height decreased the damage to vegetation, and the surviving trees further decreased the washout region of houses. However, further increasing the embankment height reduced the water overflow, and the vegetation effect decreased again because of the reduced resistance from coastal trees (Fig. 2a). A similar effect was demonstrated with changing tsunami height. Fig. 2b shows the variation of the effects with changing tsunami water depth from 7.1 to 9.7 m (at GEJT). With a smaller tsunami height than that of the GEJT, the vegetation effect becomes maximum. The study indicated that an optimal combination of embankment and forest can be determined by the tsunami extent. Therefore, a hybrid defense system (HDS) should be optimally designed not only to protect coastal communities by an embankment in a

Level 1 tsunami but to mitigate the destruction of houses by combining a coastal forest with embankment for a Level 2 tsunami.

4. Effective hybrid structures and their arrangement for tsunami mitigation

After the GEJT, coastal defense systems were strengthened in Hokkaido Prefecture to mitigate a tsunami inundation by using existing coastal forests behind which communities are assumed to be vulnerable to damage (Fig. 3a). As the extent of a future Level 2 tsunami is quite large, most of the trees will be broken in Shiranuka and Taiki Town in Hokkaido (Tanaka et al., 2018). Because it is not feasible to increase only the forest length as a mitigation measure, a hybrid structure combining the forest with an embankment and/or a moat was proposed. In order to adapt the hybrid defense system (HDS) for a tsunami, the optimal HDS for the designated area needs to be elucidated. The method to increase the mitigation function of the existing forest needs to be optimized in relation to the available land pattern, i.e., seaward, landward, or both sides of the forest (Fig. 3b). Zaha

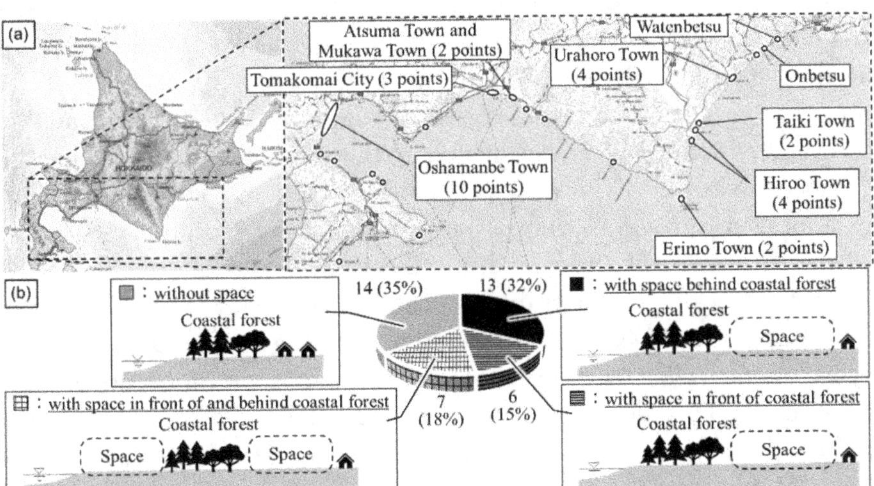

Fig. 3. Future tsunami-affected coastal forest conditions in Hokkaido. (a) Location of the affected area behind existing coastal forest (modified from GSI maps (https://maps.gsi.go.jp/) by the Geospatial Information Authority of Japan), (b) number and percentage of forests considering the available land around a forest (modified from Zaha et al. (2019)).

et al. (2019) conducted flume experiments with a physical model using three components, a coastal forest, a moat, and an embankment under a surge type flow by changing the arrangement of each component. The mitigation effects of each arrangement were quantified using indices of the reflection, transmission, fluid force reduction, delay in tsunami arrival, inundation volume, and vegetation resistance characteristics in an emergent vegetation condition. The case for submerged vegetation was further tested by Kimiwada et al. (2020), and the superiority of the HDS that Zaha et al. (2019) proposed was also confirmed. The results of Zaha et al. (2019) and Kimiwada et al. (2020) are explained in Section 4.1. Tanaka et al. (2021) further quantified the effectiveness of the HDS (Zaha et al., 2019) in a numerical simulation by changing the wavelengths of the surge-type flow. This is explained in Section 4.2.

4.1. *Optimal arrangement of vegetation (V), embankment (E), and moat (M) in surge-type flow experiment*

4.1.1. *Physical model experiment*

This chapter introduces the important points revealed in Zaha et al. (2019) and Kimiwada et al. (2020). A surge-type flow was generated by quickly lifting a gate in a flume with an HDS model under the scale of 1/100 (Fig. 4). Froude numbers were set at 1.55, 1.79, and 2.03, and the generated wave period was estimated as 150 sec on a real scale using the Froude similarity. A GPS wave gauge off the coast of Kuji recorded eight separate tsunami waves with periods of around 200 seconds for around 28 min in the GEJT (Tanaka et al., 2016). Although the period in the experiment is slightly short for a tsunami, it corresponds to the separate tsunamis at the GEJT. The study aimed to distinguish the differences of hybrid defense systems (HDSs) for mitigating the inundating flow behavior by the first wave mainly for evacuation purposes. To replicate 5–10 m tsunamis on a real scale, the model scale was decided and the tank water depth was set so that the approaching wave height became around 1–2 times the embankment height.

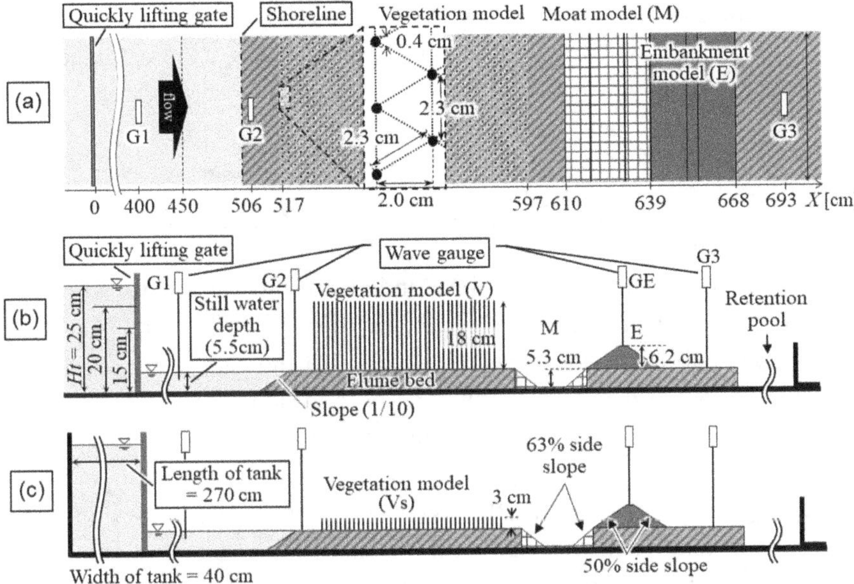

Fig. 4. Experimental setup for hybrid defense systems using a surge-type flow. (a) Plan view of one example (Case VME), (b) and (c) side view for emergent vegetation (V) and submerged vegetation (Vs) (modified from Zaha et al. (2019) and Kimiwada et al. (2020)).

An embankment and a moat model made of wooden boards were used as artificial structures. As a typical tree of the coastal forest, black pine was modelled by a circular cylinder, of which the diameter, height, density, and spacing used were scaled down to 0.4 cm, 18 cm, 0.22 cylinders/cm^2, and 2.3 cm, respectively.

For quantifying the tsunami mitigation functions of the different HDSs, the arrangements of the vegetation (V), embankment (E), moat (M), and no system (N) were adjusted. Two types of experiments were conducted for vegetation, emergent vegetation (V), and submerged vegetation (Vs), by Zaha et al. (2019) (Fig. 4(b)) and Kimiwada et al. (2020) (Fig. 4(c)), respectively. The case name represents the order of the model components from seaward.

For clarifying the differences of the investigated HDS, delay in the tsunami arrival time (AT), which is defined an elapsed time at G3 after the arrival at G1, overflow volume from the embankment ($\triangle Q$), and the reflected wave from the model were measured. In addition, the maximum reduction rates of the fluid force index and the moment index in the inland area were defined as *RFI* and *RMI* respectively. They were calculated by Eq. 3 and Eq. 4.

$$RFI = ((u^2h)_{NNN} - (u^2h)_{max})/(u^2h)_{NNN} \tag{3}$$

$$RMI = ((u^2h^2)_{NNN} - (u^2h^2)_{max})/(u^2h^2)_{NNN} \tag{4}$$

where u^2h and u^2h^2 are the fluid force index and moment index when u is the velocity and h is the water depth; $(u^2h)_{NNN}$, $(u^2h^2)_{NNN}$, $(u^2h)_{max}$, and $(u^2h^2)_{max}$ are defined as the maximum values of the fluid force index and moment index, where the subscripts 'NNN' and 'max' represent the maximum value in Case NNN (without a model), and that in each model case, respectively.

For discussing the total effectiveness of each hybrid system with different wave heights, four evaluation indices, $E_{\Delta Q}$, E_{AT}, E_{RFI}, and E_{RMI}, were defined. These are the averaged value of each index over average wave height.

4.1.2. *Physical model experiment*

Table 1 shows the effectiveness of HDS as evaluated by Zaha et al. (2019) and Kimiwada et al. (2020) using the four evaluation indices. A double circle and circle indicate the advantageous structures in that order. For the seaward vegetation type, Case VME has the highest efficacy on $E_{\Delta Q}$ and E_{AT}. For the landward vegetation type, Case EMV is judged as best from E_{RFI}. A good option when the growth of seaward coastal trees is not good due to wind-blown salt sprays is to replace the unhealthy trees with an embankment and moat to strengthen the tsunami mitigation function (Zaha et al., 2019).

When the tsunami greatly exceeds the tree heights (submerged vegetation (Vs) cases in Table 1), the order of the effectiveness changed slightly (Kimiwada et al., 2020). Even in that case, EMVs and VsME types can maintain the effectiveness of landward and seaward vegetation types, respectively, although the value becomes smaller compared with the emergent vegetation case.

Table 1. Differences in mitigation effect of each HDS through four indices (modified from Zaha et al. (2019) and Kimiwada et al. (2020))

Case	Vegetation model	$E_{\Delta Q}{}^{*}$		$E_{AT}{}^{**}$		$E_{RFI}{}^{***}$		$E_{RMI}{}^{****}$	
VME	Emerged	89.8	◎	2.1	◎	74.3		92.4	◎
VEM	Emerged	83.9		1.9	○	75.5		84.8	
EMV	Emerged	65.5		1.1		83.1	◎	90.2	○
MEV	Emerged	65.4		1.2		77.3	○	84.6	
MVE	Emerged	86.8	○	1.7		59.2		84.2	
EVM	Emerged	61.5		1.4		72.4		79.4	
VsME	Submerged	77.2	◎	0.7		40.9		71.6	
VsEM	Submerged	70.5		0.7		23.6		23.6	
EMVs	Submerged	61.1		0.9	◎	68.7	◎	80.8	◎
MEVs	Submerged	65.1		0.9	◎	56.7		69.5	
MVsE	Submerged	76.4	○	0.9	◎	37.0		66.4	
EVsM	Submerged	60.0		0.9	◎	66.7	○	76.4	○

$E_{\Delta Q}{}^{*}$, $E_{AT}{}^{**}$, $E_{RFI}{}^{***}$, $E_{RMI}{}^{****}$ are defined as the averaged values of overflow rate, arrival time, *RFI*, *RMI* respectively over average wave height.

Even in emergent vegetation cases, the tsunami current can easily pass through the top layer of vegetation when overturning occurs (Igarashi and Tanaka, 2018). The effective HDS summarized in Table 1 may be changed in that case. For this situation, more study is needed to clarify the effect by numerical simulations.

Although Case VME is the best for the seaward vegetation type and has large tsunami mitigation effect, the forest itself can be a little weak if the first push wave is large. However, the disadvantage of producing driftwood was small when the fluid force reduction was large, like the case in the Sendai Plain at the GEJT (Tanaka and Ogino, 2017). In an actual case, large diameter trees for trapping debris should be provided at least on the landward side within the seaward vegetation. Within the landward vegetation type, Case EMV was optimal to reduce the fluid force on the coastal forest because energy was reduced in front of the forest by a hydraulic jump. The EMV system itself can reduce the destruction of the trees except for the frontal side of the forest, and thus can increase not only the vegetative resistance but additional resistance by trapping driftwood (Tanaka and Onai, 2017, Rahman et al., 2022).

More experiments are needed considering the change in vegetation length in HDS and the combination of the size of E and M. The tree breaking effect (Fig. 1) in HDS should be elucidated in general by numerical simulations like those of Tanaka et al. (2018).

4.2. *Effect of changes in tsunami wave length on tsunami mitigation by hybrid structures VME and EMV*

Although Level 2 tsunamis at the coasts in Hokkaido are estimated to be around 2–35 m (HFS, 2016) and tsunami periods usually range from several minutes to longer than one hour, the experiments by Zaha et al. (2019) and Kimiwada et al. (2020) did not cover the full range of the Level 2 tsunami cases. As a flow around the peak of long wave-like tsunami is usually considered to be a quasi-steady state, steady state experiments were also conducted for evaluating the effectiveness [Ali and Tanaka, 2019; Rahman et al., 2021]. However, as HDS has large advantages in delaying the tsunami arrival time and reducing the overflow volume, the unsteady effect should be discussed for even longer tsunami period and wavelength cases. Even though these studies (Zaha et al., 2019; Kimiwada et al., 2020) conducted unsteady state experiments, the range was a little small. Thus, a numerical simulation with a validated model is very useful for understanding the real situation by changing the wave period.

Tanaka et al. (2021) changed the approaching tsunami conditions (wave length, wave period, and wave height) and demonstrated the change in the mitigation effects by the HDS proposed by Zaha et al. (2019). Simulation conditions (200–580 seconds using Froude similarity) were set by numerically increasing the length of the water tank used in Zaha et al. (2019).

Using the indices defined in Section 4.1, the change of effectiveness was simulated. The indices show the superiority of the HDS in shorter tsunami periods; however, the smaller mitigation effect can be retained even in the long period cases. Among the long period cases, Case VME was advantageous for controlling the overflow volume from the embankment, the delay in tsunami arrival time, and the reduction of moment index (RMI) in Eq. 4. Case EMV was advantageous to reduce the fluid force index (RFI) in Eq. 3. The numerical simulation validated the knowledge of physical experiments by Zaha et al. (2019).

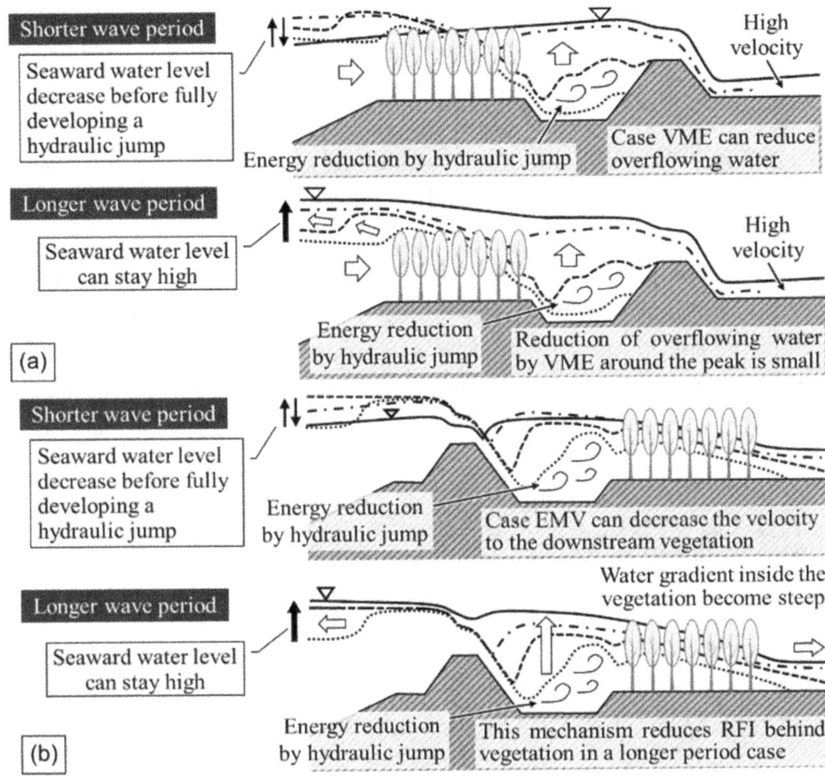

Fig. 5. Schematic of the differences in hybrid structures with the change in tsunami period. (a) VME, (b) EMV (modified from Tanaka et al. (2021))

Case VME increased the water level in front of the embankment by generating a hydraulic jump in between vegetation and embankment (Fig. 5a). For the shorter wave period, the water level at seaward starts to decrease before fully developing a hydraulic jump and the stored water in the region moves seaward. Case VME can thus decrease the overflowing water from the embankment. For a longer tsunami period, the seaward water level can remain higher than the water level of the hydraulic jump after reaching a quasi-steady state. Then, the reduction of overflowing water around the peak decreases in comparison with the shorter tsunami period case.

Case EMV can decrease the velocity to the downstream of vegetation. In a shorter tsunami period, the seaward water level starts to decrease before the fully developed hydraulic jump, although the overflow from the embankment is not affected by the inland-side MV system. On the other hand, a fully developed hydraulic jump in a longer period case can further raise the water level around the moat and increase the velocity just downstream of the vegetation due to a steep water gradient inside the vegetation. This mechanism reduced RFI behind the vegetation in a longer period case (Fig. 5b).

Due to the dispersibility and nonlinearity of a tsunami, soliton fission can make short waves (Baba et al., 2015). A storm surge can also make surf beats, relatively short waves (Roeber and Bricker, 2015). The usefulness of HDS can be assumed to increase for the short waves included in the long wave because it can directly affect the hydrodynamics around the peak.

5. Installation of two types of hybrid structures in Hokkaido Pref., Japan

As explained in Section 4, defense systems that strengthen the disaster mitigation function of existing coastal forests for a future Level 2 tsunami by combining the forest with an embankment and/or a moat (HDS) were proposed for the Watenbetsu district in Shiranuka Town and Onbetsu district in Kushiro City, Hokkaido Prefecture (Fig. 3a). Two hybrid defense systems (HDSs) were selected based on the existing forest and land availability. This section describes the HDS installed in the two locations (Fig. 6) and evaluates the mitigation effect of Level 2 at the Onbetsu site. As discussed by Zaha et al. (2019), one of the most important advantages of the EMV for a huge Level 2 tsunami is to delay the tsunami arrival time. Thus, the main advantage is mainly explained by how the EMV changes the evacuation vulnerability.

5.1. *Pilot project installing VME (Watenbetsu, Shiranuka Town) and EMV (Onbetsu, Kushiro City)*

Hybrid defense systems (HDSs), VME and EMV, have been actually installed in Watenbetsu district in Shiranuka Town and Onbetsu district in Kushiro City, respectively, in Hokkaido Pref., as shown in Fig. 6.

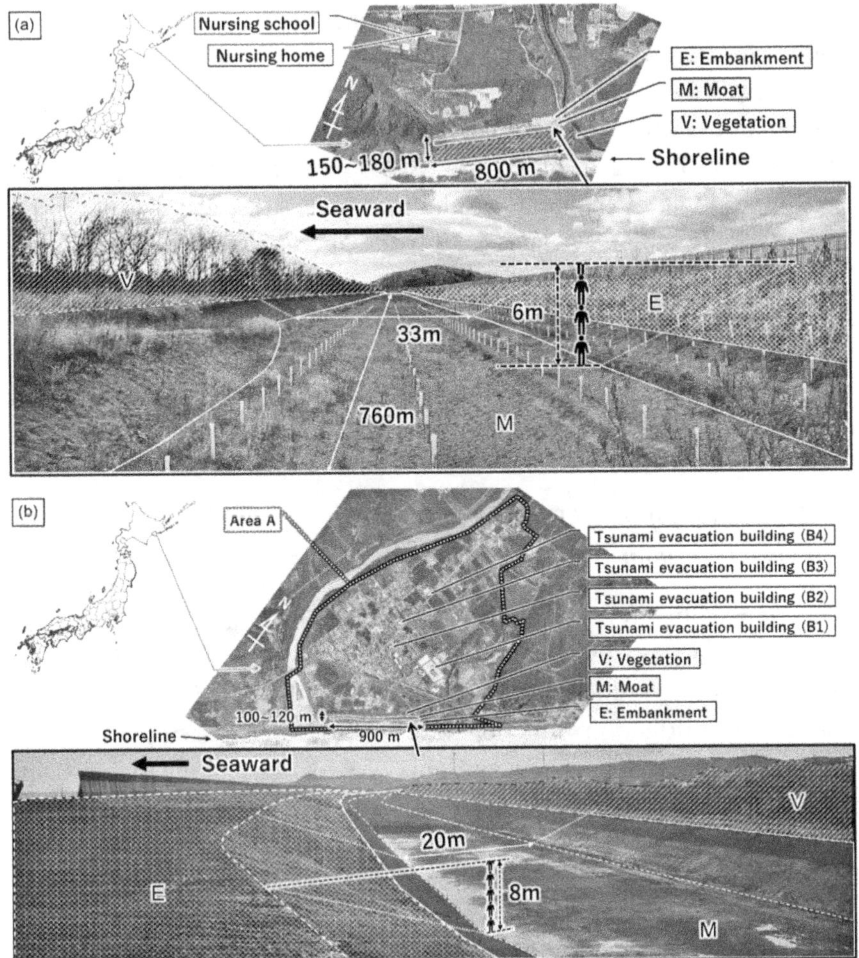

Fig. 6. Installed hybrid structures. (a) VME system: Watenbetsu (Shiranuka Town), (b) EMV system: Onbetsu (Kushiro City). (Aerial photo provided by the Geospatial Information Authority of Japan).

In the Watenbetsu district, there were 155 households (461 people) in 2014, and there were a nursing home and nursing school behind the planned HDS. The growth of the coastal vegetation itself was healthy at this point. The objective of installing the VME-type HDS was to reduce the inundating water volume and increase the time for evacuation by delaying the tsunami arrival time by utilizing the existing seaward forest.

The Onbetsu district (Area A in Fig. 6b) had 838 households (1356 people) in 2022. At this location, many wooden buildings existed just behind the planned HDS. The growth of the coastal vegetation itself was not good due to the strong wind and salt spray at this location. The objective of installing the EMV-type HDS was to replace the seaward vegetation of the embankment (E) and moat (M) to reduce the fluid force and increase the time for evacuation by delaying the tsunami arrival time. There are four tentative tsunami-evacuation buildings at this site. However, because the extent of a Level 2 tsunami is quite large, the evacuation buildings are not perfectly safe. Not only for the tentative evacuation but also for the evacuation to the outside of the inundation area, it is very important to delay the tsunami arrival time at this site.

5.2. *Evaluation of hybrid structure in Onbetsu (EMV) for Level 2 tsunami*

This section evaluates the effectiveness of HDS in Onbetsu (EMV system) in a Level 2 tsunami by a numerical simulation. The method is almost identical to that described in Tanaka et al. (2018), where it was applied to evaluate the mitigation function of the existing coastal forests in Watenbetsu in Shiranuka Town and Taiki Town, close to the Onbetsu site in Hokkaido. (For details of the numerical model, see Tanaka et al. (2018)).

5.2.1. *Numerical simulation model*

Similar to Tanaka et al. (2018), five regions, Regions A, B, C, D, and E, of which grid sizes were set to 1350 m, 450 m, 150 m, 50 m, and 16.7 m, respectively, were used to calculate the tsunami propagation (Fig. 7a). Using a nesting method, the simulated values in a large region were applied in the next smallest region. Linear long-wave equations and non-linear long-wave equations were applied to Region A and Regions B–E, respectively. At the coast, a perfect reflection boundary was applied in Regions A–C, and inundation was calculated for Regions D and E.

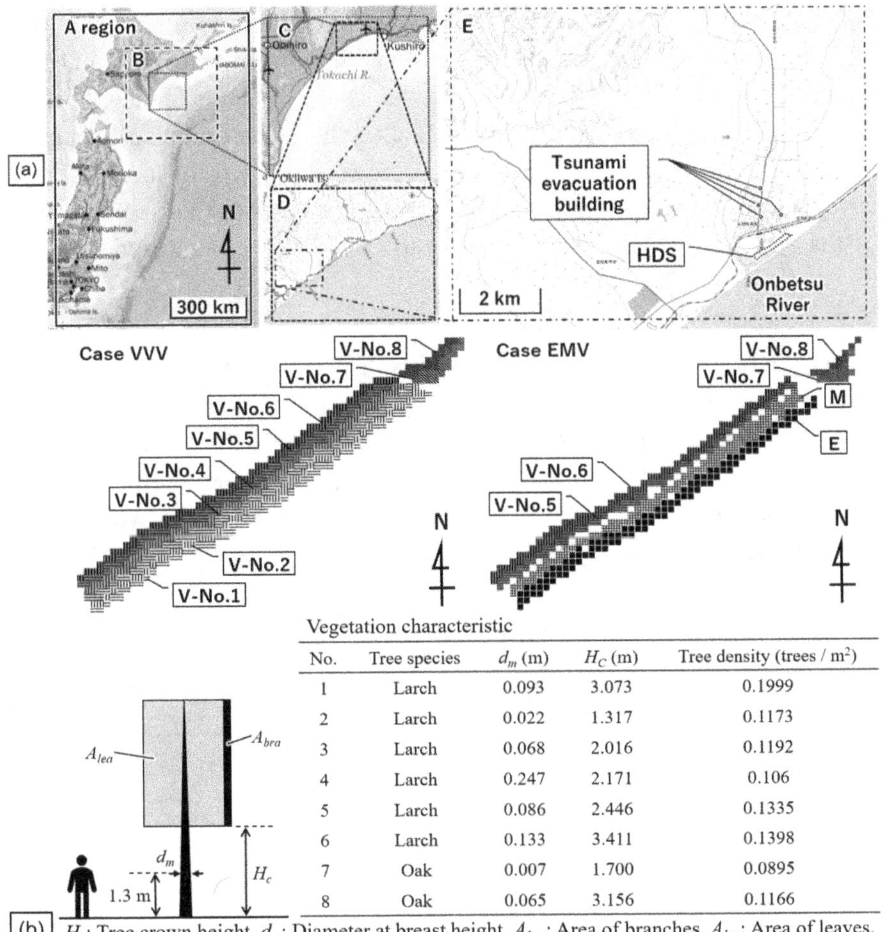

Vegetation characteristic

No.	Tree species	d_m (m)	H_C (m)	Tree density (trees / m²)
1	Larch	0.093	3.073	0.1999
2	Larch	0.022	1.317	0.1173
3	Larch	0.068	2.016	0.1192
4	Larch	0.247	2.171	0.106
5	Larch	0.086	2.446	0.1335
6	Larch	0.133	3.411	0.1398
7	Oak	0.007	1.700	0.0895
8	Oak	0.065	3.156	0.1166

H_c: Tree crown height, d_m: Diameter at breast height, A_{bra}: Area of branches, A_{lea}: Area of leaves.

Fig. 7. Definition of the grid system and the model used. A: Linear long-wave equations, B–D: non-linear long-wave equations, E: non-linear long-wave equations with a turbulence model. (Aerial photo provided by the Geospatial Information Authority of Japan).

To estimate resistance by surface roughness, a Manning roughness coefficient (n) that doesn't include the resistance by vegetation was set based on the land utilization, i.e., 0.025, 0.03, 0.02, 0.04, 0.06, and 0.08 s/m$^{1/3}$ for sea, coastal forest, paddy field, wetland, residential area (medium density), and residential area (high density), respectively. For drag resistance by vegetation, the same model was applied as described in Tanaka et al. (2018). Fig. 7b shows the modeled tree characteristics. The resistance by vegetation change considering the tree breaking was included in 2.2 (Fig. 1) using the M_{criX} and M_{criOT} (Eqs. 1 and 2) where σ_{MAX} is 26.3 MPa and 17.6 MPa for larch and oak in Onbetsu Dist., and (k_O k_C) are the same values as those in Shiranuka Town for larch and oak described in Tanaka et al. (2018). For each parameter of tree species given by the allometry equation measured at the sites below were used.

$$y = k\,x^c \tag{5}$$

where k and c are dimensional constants. Table 2 shows the coefficients of the trees in Onbetsu Dist.

Sea surface displacements were applied using the fault model of Mansinha and Smylie (1971) as initial conditions of the simulation. The fault parameter for a Level 2 tsunami in the Hokkaido area was applied from the published data (DPH, 2012).

Table 2. Coefficients of eqs. 1, 2 and 5

Tree species (Location)	Allometry equation measured at Onbetsu				Coefficient related to tree breaking		
	y	x	k	c	σ_{MAX} (MPa)	k_O	k_C
Larch	H	d_m	7.6984	0.2754	26.3	26.0	3.1
(Onbetsu)	A_{bra}	$d_m^2 H_t$	10.922	0.7112			
	A_{lea}	$d_m^2 H_t$	29.642	0.67			
Oak	H	d_m	15.757	0.4298	17.6	86.4	0
(Onbetsu)	A_{bra}	$d_m^2 H_t$	7.1959	0.7458			
	A_{lea}	$d_m^2 H_t$	139.55	0.67			

H_t: Tree height, d_m: Diameter at breast height, A_{bra}: Area of branches, A_{lea}: Area of leaves.

5.2.2. *Effects of hybrid structure for mitigating the inundation of Level 2 tsunami (recurrent period: several hundred– thousand years)*

Fig. 8 shows the simulated tsunami inundation results for Case VVV (existing coastal forest condition) and Case EMV at Onbetsu. Fig. 8(a) shows the maximum inundation depth (h_{max}) for Case EMV. Because a Level 2 tsunami is quite high, a large area is affected by the tsunami inundation although one of the HDS, EMV system, is installed. Fig. 8(b) shows the difference of the maximum water depths from Case EMV to Case VVV. Minus values show the advantage of installing the EMV system on the water depth decrement.

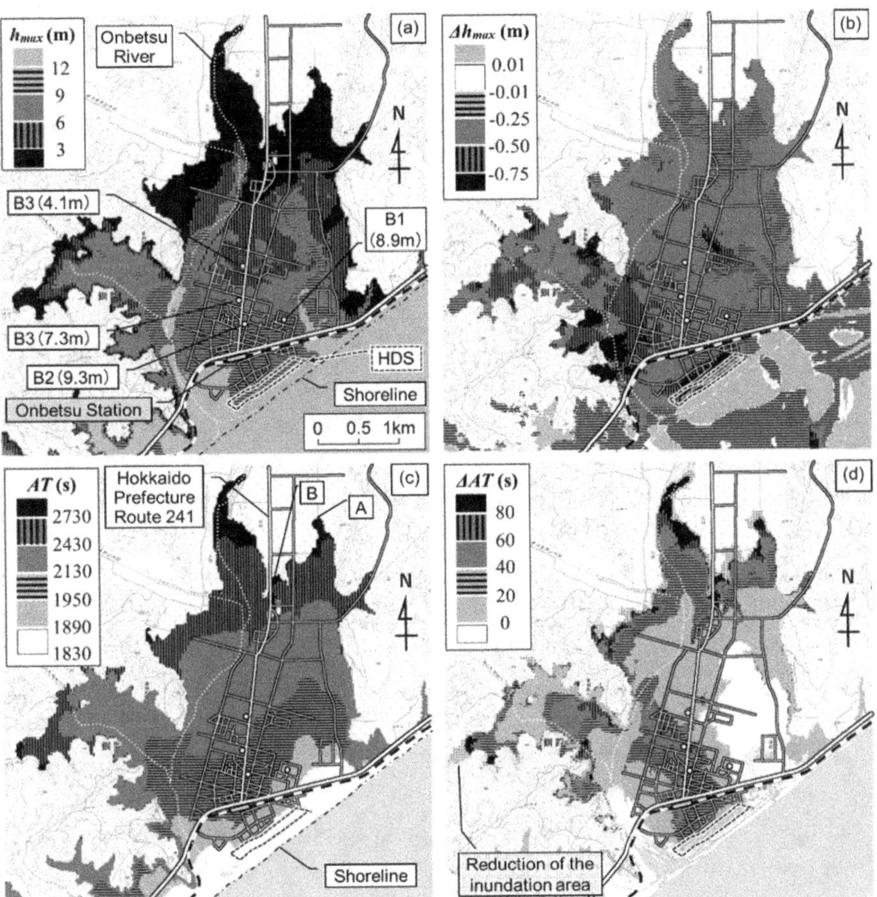

Fig. 8. Simulated tsunami inundation at Onbetsu. (a) Maximum inundation depth for Case EMV, (b) the difference of the maximum water depth (Case EMV − Case VVV), (c) tsunami arrival time (AT) for Case EMV, and (d) the difference of AT in two cases ($\Delta AT=AT$ in Case EMV−AT in Case VVV). (Aerial photo provided by the Geospatial Information Authority of Japan).

Fig. 8(c) shows the tsunami arrival time (AT) in Case EMV. After the fault motion started, the tsunami arrived at the shoreline after 1830 seconds. At 3, 23, and 29 minutes after the inundation started, Onbetsu station, Location A, and Location B were inundated, respectively. By introducing EMV, the inundated area became smaller, and the AT was delayed around 1 minute. This contributes to evacuation outside the inundated area.

Fig. 8(d) shows the effectiveness of using Road 241 for evacuation from the inundated area. Because Road 241 is located behind the EMV system, the delay in tsunami arrival time can be expected to be large. In contrast, there are rivers on the east and west sides of the road that will propagate the tsunami faster than the inland Road 241 area. Then, AT at Location B was 6 minutes later that that at Location A. To utilize Road 241 as an emergency evacuation route, it is also important to plant trees along the road for trapping people, providing soft landings, and climbing, which are important functions of coastal forests learned from post tsunami surveys (Section 2.1).

The distance from Onbetsu Station to the outside of the inundated area is around 3 km. If we assume the walking speed is 1 m/s, it takes 50 minutes to get out of the inundated area. Then, if the evacuees started walking after the arrival of tsunami at the coast, the evacuees cannot go outside. There are four evacuation buildings in the Onbetsu district. The water depth decrement due to the EMS system at the emergency evacuation buildings is low (around 23–48 cm), as shown in Fig. 8(b). The EMV system cannot increase the possibility of using the lower floors of the buildings, although EMV can increase the time available for evacuation to the building. Therefore, the evacuees need to start evacuation within several minutes of the fault motion to escape the inundated area. In that case, the delay in tsunami arrival time is quite important for peoples' survival.

Fig. 9 shows the simulated damage pattern of the coastal forest at Onbetsu. In the existing coastal forest case, Case VVV, tree trunk breakage or overturning occurs just after the tsunami arrival inland (elapsed time: 1843–1868 seconds after the fault motion starts). Tree trunk breakage occurs in a wide area along the coast, and behind the trunk breakage, overturning occurs on the inland side. Thus, the tsunami can easily pass over the forest and the trapping effect by the inland-side forest cannot be expected. On the contrary, in Case EMV, the tree damage in most of the locations shows the overturning pattern, which does not produce large pieces of driftwood, even though tree trunk breakage also occurs along the Onbetsu River side. In addition, the elapsed time when the tree trunk breakage occurs is around 1990–2073 seconds, 2–4 minutes later than the Case VVV. It is helpful from the point of view of people evacuation because the possibility of the secondary damage by driftwoods decreases.

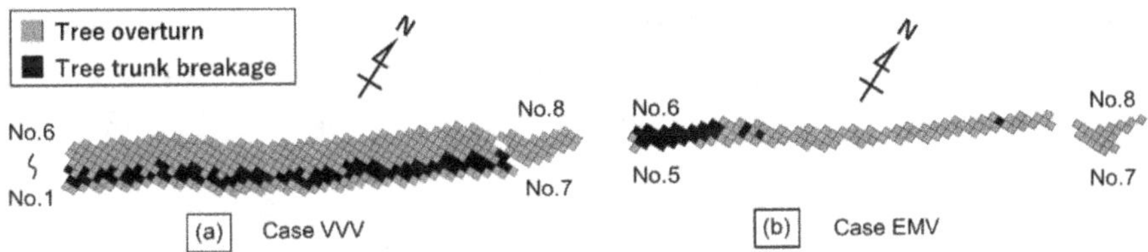

Fig. 9. Simulated damage pattern of coastal forest at Onbetsu for (a) previous coastal forest (VVV) and (b) installed EMV system

6. Change in tsunami mitigation function by forest management

This section discusses how tree growth and forest thinning affect the tsunami mitigation effect using non-dimensional run-up height (NR = run-up height with vegetation (R)/run-up height without vegetation (R0)) using the model explained in 5.2.2. Related to the management of a tsunami-mitigation forest, Igarashi et al. (2019) demonstrated the mitigation effects by changing tsunami heights (5–20 m) and forest width (10–640 m) for Dahurian larch under a thinning regime. Torita et al. (2022) also analyzed the effects of three thinning conditions of Japanese black pine (*Pinus thunbergii* Parlat.) for a 100-m forest. This chapter introduces 300-m forest cases under the same thinning, topography (slope, roughness), and tsunami conditions.

Fig. 10a shows a schematic of the boundary conditions of a tsunami, location of the vegetation model, and topography. For the boundary conditions, the time series of water level and flow velocity were given at the boundary (at the place with the still water depth 150 m) to generate a sine wave (wave period: 1800 s) so that the maximum flow depths at the shoreline become 5 m, 10 m, and 15 m, respectively. Fig. 10b shows the forest characteristics (tree height, trunk diameter, tree-crown height, and tree density) for the two thinning regime conditions observed in Hokkaido Pref. DT is the case where no artificial thinning was conducted, but not-well grown trees were eliminated by natural selection, and the tree density became gradually smaller. On the other hand, the tree density of moderate thinning (MT) decreased due to periodic artificial thinning. With the MT treatment, the forest tends to larger trunk diameter and lower tree crown height than with DT. The model parameters for tree-breaking conditions and other parameters for estimating the tree resistance are calculated using the allometry equations and tree data shown in Fig. 10b. The equations and the coefficients are the same as those reported by Torita et al. (2022).

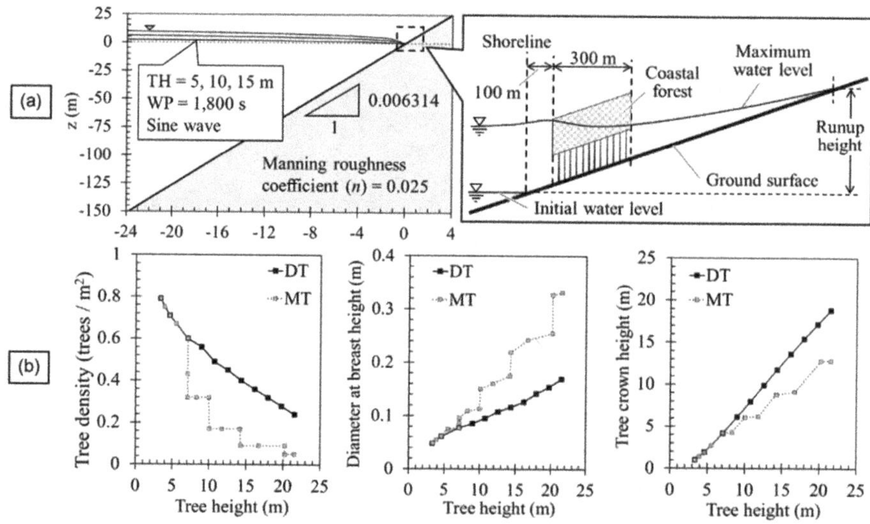

Fig. 10. Numerical simulation conditions. (a) Topography and tsunami conditions, (b) vegetation conditions for two thinning regimes (DT and MT).

First, the growth of DT without thinning treatment is explained. Fig. 11 shows the relationship between tree height and non-dimensional run-up height. Three tsunami heights at the coast, 5, 10, and 15 m, are compared. Three type plots, white, grey, and black, indicate that the major destruction types are standing (not broken), overturning, and tree-trunk breakage, respectively. It shows the smaller the NR, the larger the mitigation effect. The NR values ranged around 0.85–0.89 for a 5-m tsunami at the coast. It means a 11–15% reduction of the inundated area for the constant slope case. The reduction becomes 5% (0.95 of NR) for a 10 m tsunami with small trees, but it increases to around 17% (0.83 of NR) with the tree growth because the trees are not broken even in the larger tsunami.

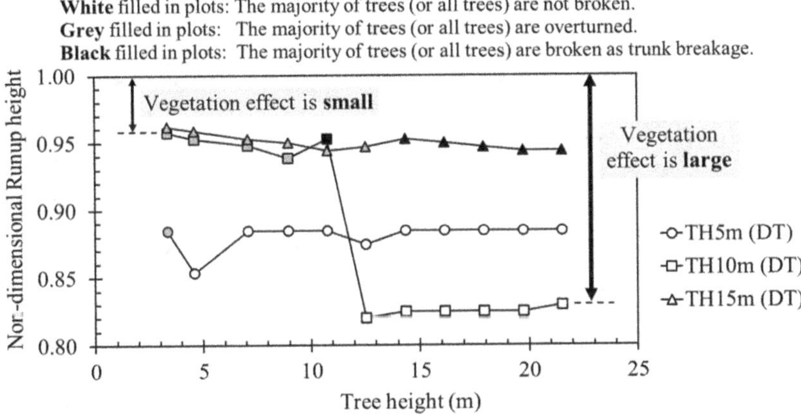

Fig. 11. Change in non-dimensional tsunami runup (NR) with respect to tree height for three tsunami heights (THs).

For understanding the effects of artificial thinning as forestry, the NR for DT was compared with that for MT under which the thinning treatments are continuously provided (Fig. 12a). The NR values are greatly changed with the increase of resistance due to tree growth, the decrease of resistance with the decrease in tree density due to thinning treatment, and the change of tree destruction mode with the tree growth. Fig. 12b shows the schematic of the change in NR. Compared with DT (without artificial thinning), the MT treatment increases the tree diameter growth and decreases the tree crown height. In most of the cases, DT shows smaller NR in comparison with MT, but the fluctuation due to the growth of trees and thinning can reverse the order (NR of MT becomes smaller than that of DT, like with a 5-m tree height and TH=10 m). In that case, MT can decrease run-up height compared with DT. However, just after MT treatment, the NR becomes larger because the tree density becomes smaller. In addition, the NR becomes smaller (mitigation effect becomes larger) when the tree destruction mode is changed from overturn or trunk breakage to standing (not affected) due to the increase of vegetative drag force. On the other hand, if the breaking mode is changed from overturning to trunk breakage, the NR becomes larger (mitigation effect becomes smaller) because of the decrease of the resistance from the washed-out part over the trunk breakage.

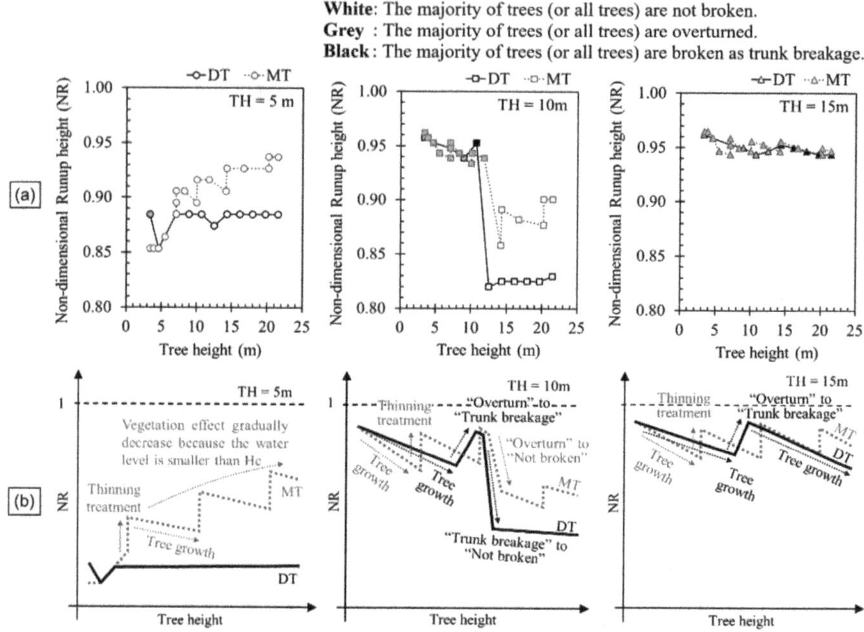

Fig. 12. Effects of thinning treatment (DT, MT) on non-dimensional tsunami run-up (NR) with respect to three tsunami heights (THs). (a) Simulated NR, (b) schematic of the changing mechanism of (a).

For every tsunami water depth at the coast, most cases of DT show smaller NR in comparison with MT. This indicates that DT is good for tsunami mitigation. However, the tree growth itself is not good as forestry because trees remain thin with a small diameter. In case of MT treatment, the thinning can increase the trunk diameter and lower the tree crown height. In that case, trees can be affected by a lower height tsunami, and overturning can occur due to a large resistance by the tree crown. On the contrary, the tree developed by DT has a smaller diameter and higher tree crown height, and thus tree trunk breakage can occur. Then, forest management should be considered not only for keeping resistance high but also for controlling the tree breaking mode as overturning (Torita et al., 2022).

7. Summary

This chapter introduced the importance of selecting tree species for a coastal forest and its management as the tsunami buffer zone and for the hybrid defense system comprising natural (coastal forest) and artificial (embankment and moat) components. A summary is shown below.

1) Considering the functional limitations (destruction of trees) and disadvantage (production of driftwood), a hybrid defense system (HDS) that strengthens existing coastal forests for a future Level 2 tsunami by combining the forest with an embankment and/or a moat has been proposed. Two HDSs, VME (vegetation, moat, and embankment from seaward) and EMV (opposite order of VME from seaward), have been installed in Watenbetsu in Shiranuka Town and Onbetsu in Kushiro City, Hokkaido Prefecture, respectively, based on the existing forest condition and land availability.

2) For the costal forest construction, stand structures of vegetation (density, diameter, tree crown height, etc.) and their critical values for breaking are important to increase the mitigation effect. Forest management (in forestry, thinning) should be based on the parameters in each species.

3) Hydraulic jump formation by HDS is an important mechanism for the reduction of tsunami inundation. Trapping debris can be utilized for energy mitigation in some cases.

Although we have conducted many studies for HDS, the model scale is a little small. When a hydraulic jump is generated, we need to consider the 'scale effect'. Larger scale experiments should be conducted in the future, and be numerically validated to generalize the effects.

References

1. T. Hiraishi and K. Harada, Greenbelt tsunami prevention in South-Pacific region. *Report of the Port and Airport Research Institute* 42(2), 23 pp. (2003).
2. F. Danielsen, M.K. Sorensen, M.F. Olwig, V. Selvam, F. Parish, N.D. Burgess, T. Hiraishi, V.M. Karunagaran, M.S. Rasmussen, L.B. Hansen, A. Quarto and N. Suryadiputra, The Asian tsunami: A protective role for coastal vegetation. *Science* 320 (5748): 643 (2005).
3. K. Kathiresan and N. Rajendran, Coastal mangrove forests mitigated tsunami. *Estuarine, Coastal and Shelf Science* 65(3), 601–606 (2005).
4. N. Tanaka, Y. Sasaki, M.I.M. Mowjood and K.B.S.N. Jinadasa, Coastal vegetation structures and their functions in tsunami protection: Experience of the recent Indian Ocean tsunami. *Landscape and Ecol. Eng.* 3, 33–45 (2007).
5. H. Yanagisawa, S. Koshimura, K. Goto, T. Miyagi, F. Imamura, A. Ruangrassamee and C. Tanavud, The reduction effects of mangrove forest on a tsunami based on field surveys at Pakarang Cape, Thailand and numerical analysis. *Estuarine, Coastal and Shelf Science* 81, 27–37 (2009)
6. N. Tanaka, S. Yasuda, K. Iimura and J. Yagisawa, Comparison of the effects of coastal forest and those of sea embankment on reducing the washout region of houses in the tsunami caused by the Great East Japan Earthquake. *J. Hydro-environment Research* 8, 270–280 (2014).
7. S. Temmerman, P. Meire, T.J. Bouma, P.M. Herman, T. Ysebaert and H.J. De Vriend, Ecosystem-based coastal defence in the face of global change. *Nature* 504(7478), 79–83 (2013).
8. A.E. Sutton-Grier, K. Wowk and H. Bamford, Future of our coasts: The potential for natural and hybrid infrastructure to enhance the resilience of our coastal communities, economies and ecosystems. *Environmental Science & Policy* 51, 137–148 (2015).
9. Y. Igarashi, N. Tanaka and T. Zaha, Changes in flow structures and energy reduction through compound tsunami mitigation system with embankment and lined piles. *Ocean Engineering* 164, 722–732 (2018).
10. Y. Igarashi and N. Tanaka, Effectiveness of a compound defense system of sea embankment and coastal forest against a tsunami. *Ocean Engineering* 151, 246–256 (2018).
11. T. Zaha, N. Tanaka and Y. Kimiwada, Flume experiments on optimal arrangement of hybrid defense system comprising an embankment, moat, and emergent vegetation to mitigate inundating tsunami current. *Ocean Engineering* 173, 45–57 (2019).
12. A.H.M. Rashedunnabi and N. Tanaka, Energy reduction of a tsunami current through a hybrid defense system comprising a sea embankment followed by a coastal forest. *Geosciences* 9, 247 (2019).
13. S. Honda, Über den Küstenshutzward gegen Springfluten. *Bulletin of the College of Agriculture* 3(4), Imperial University, Komaba, Tokyo, pp. 281–298 (1898).

14. N. Shuto, The effectiveness and limit of tsunami control forests. *Coastal Eng. Jpn.* 30 (1), 143–153 (1987).

15. L. Dengler and J. Preuss, Mitigation lessons from the July 17, 1998 Papua New Guinea tsunami. *Pure and Applied Geophysics* 160, 2001–2031 (2003).

16. Food and Agriculture Organization of the United Nations (FAO). The Role of Coastal Forests in the Mitigation of Tsunami Impacts – Main report" www.fao.org/forestry/site/coastalprotection/en. ISBN 978-974-13-9321-3, *Thammada Press* (2007).

17. N. Tanaka, J. Yagisawa and S. Yasuda, Breaking pattern and critical breaking condition of Japanese pine trees on coastal sand dunes in huge tsunami caused by Great East Japan Earthquake. *Nat. Hazards* 65, 423–442 (2013).

18. A. Ali and N. Tanaka, Experimental study of scouring downstream of coastal vegetation in an inundating tsunami current. *Landscape Ecol. Eng.* 16(4), 273–287 (2020).

19. N. Tanaka and A. Onai, Mitigation of destructive fluid force on buildings due to trapping of floating debris by coastal forest during the Great East Japan tsunami. *Landscape Ecol. Eng.* 13, 131–144 (2017).

20. N. Tanaka and T. Ogino, Comparison of reduction of tsunami fluid force and additional force due to impact and accumulating after collision of tsunami-produced driftwood from a coastal forest with houses during the Great East Japan tsunami. *Landscape Ecol. Eng.* 13(2), 287–304 (2017).

21. M.A. Rahman, N. Tanaka and N. Anjum, Damming effects of tsunami-borne washed-out trees in reducing local scouring and tsunami energy behind a coastal embankment. *Applied Ocean Research* 126, 103260 (2022).

22. N. Tanaka, H. Sato, Y. Igarashi, Y. Kimiwada, H. Torita, Effective tree distribution and stand structures in a forest for tsunami mitigation considering the different tree-breaking patterns of tree species. *J. of Environmental Management* 223, 925–935 (2018).

23. M.B. Samarakoon, N. Tanaka and J. Yagisawa, Effects of local scouring and saturation of soil due to flooding on maximum resistive bending moment for overturning *Robinia pseudoacacia. Landscape Ecol. Eng.* 9, 11–25 (2013).

24. N. Tanaka, Y. Niwata, S. Sato, H. Torita and H. Noguchi, High-precision tsunami simulation method considering the difference in breaking and overtopping phenomenon by stand structure of coastal trees applicable to coastal forest management (in Japanese with English abstract), *J. Japan Society of Civil Engineers, Ser. B2 (Coastal Engineering)* 71(2), I_307-I_312. (2015).

25. Y. Kimiwada, N. Tanaka and T. Zaha, Differences in effectiveness of a hybrid tsunami defense system comprising an embankment, moat, and forest in submerged, emergent, or combined conditions. *Ocean Engineering* 208, 107457 (2020).

26. N. Tanaka, Y. Igarashi and T. Zaha, Numerical investigation of the effectiveness of vegetation-embankment hybrid structures for tsunami mitigation introduced after the 2011 tsunami. *Geosciences* 11(11), 440 (2021).

27. M.A. Rahman, N. Tanaka and A.H.M. Rashedunnabi, Flume experiments on flow analysis and energy reduction through a compound tsunami mitigation system with a

seaward embankment and landward vegetation over a mound. *Geosciences* 11, 90. (2021).

28. Hokkaido Research Organization, Forestry and Forest Products Research Institute, and Saitama University (HFS). The Important Research Report on the development of the method to control and manage a coastal forest for reducing tsunami energy (in Japanese), 69 pp. (2016).

29. T. Baba, N. Takahashi, Y. Kaneda, K. Ando, D. Matsuoka and T. Kato, Parallel implementation of dispersive tsunami wave modeling with a nesting algorithm for the 2011 Tohoku tsunami. *Pure Appl. Geophys.* 172, 3455–3472 (2015).

30. V. Roeber and J.D. Bricker, Destructive tsunami-like wave generated by surf beat over a coral reef during Typhoon Haiyan. *Nat. Commun* 6, 1–9 (2015).

31. L. Mansinha and D.E. Smylie, The displacement field of inclined faults. *Bulletin of the Seismological Society of America* 61(5), 1433–1440 (1971).

32. Disaster Prevention Group in Hokkaido Bureau (DPH). *Report on tsunami inundation area along Pacific Coast in Hokkaido* (in Japanese), 57 pp (2012).

33. Y. Igarashi, N. Tanaka, H. Sato and H. Torita. Tsunami mitigation effect and tree breaking situation of *Dahurian larch* coastal forest at six growth stages under thinning management of trees. In Proceedings of the 38th IAHR world congress, Panama City, Panama (2019).

34. H. Torita, Y. Igarashi and N. Tanaka, Effective management of Japanese black pine (*Pinus thunbergii* Parlat.) coastal forests considering tsunami mitigation. *J. Environmental Management* 311, 114754 (2022).

Chapter 6

The Role of Vegetation Flexibility in Wave Attenuation

T.J. van Veelen*, D.E. Reeve and H. Karunarathna

*Swansea University, Zienkiewicz Centre for Computational Engineering,
Bay Campus, Swansea, United Kingdom*

*University of Twente, Water Engineering and Management, Enschede,
The Netherlands*

**thomas.vanveelen@swansea.ac.uk, now: t.j.vanveelen@utwente.nl*

Flexible coastal vegetation such as salt marsh grasses and seagrass blades can dampen incoming waves, which benefits coastal protection. Flexible vegetation bends and sways under wave forcing. These dynamics lower the capacity of flexible vegetation to dampen waves in comparison to rigid vegetation. Solving the reduction of wave damping capacity remains an engineering challenge as it depends on the applicable wave-vegetation interaction regime. We identify six wave-vegetation regimes that depend on wave and vegetation characteristics. The non-buoyant vegetation regimes can be classified based on the Cauchy number and the excursion ratio. Under the common quasi-flexible vegetation regime, wave damping can be solved using a cantilever beam model. The reduction of wave damping by flexible vegetation compared to rigid vegetation can be quantified using a so-called work factor, which depends on the wave and vegetation properties. The solution extends towards the small tilt and rigid regimes, but not to the fully flexible regime. The analysis and model presented here provide a tool that can be applied to solve wave attenuation over flexible coastal vegetation.

1. Introduction

The motion of flexible vegetation can significantly alter the wave damping dynamics of coastal vegetation. Vegetation can be classified as flexible when its rigidity is insufficient to withstand the forces exerted by waves, such that it will bend and sway over a wave cycle.[1] In turn, the motion of plants affects the dynamics of incoming waves, creating a two-way interaction between waves and vegetation that cannot be found in rigid vegetation fields.[2-5] Common flexible coastal vegetation types include thin grasses found on salt marshes, kelp, and the blade-like vegetation of seagrass meadows.[1,2,6-11]

The wave damping capacity of flexible vegetation is less than its rigid counterpart. Field observations[10,12-15] and large-scale wave flume experiments[16-20] have shown the capacity of kelp, salt marshes and seagrass to attenuate waves, which is beneficial to coastal protection.[21-23] However, both physical experiments[24-26] and numerical modelling simulations[27-29] have shown that flexibility reduces the wave attenuation capacity of vegetation fields. For salt marsh grasses, it has been shown that a flexible grass may be up to 70% less effective in attenuating waves than a rigid stem of equal dimensions.[24,26] The exact loss in wave damping efficiency depends on wave conditions and vegetation characteristics.[4,5,24,30,31]

The two-way interaction between waves and flexible vegetation is the main driver for the loss in wave damping capacity. The bending of flexible vegetation reduces the frontal area of a vegetation stem exposed to the waves[25] and the direction of the wave forces.[30,32] Additionally, the swaying of the vegetation reduces the relative velocity between waves and vegetation, which lowers the work done by the wave forces on the vegetation and, consequently, the energy dissipation over vegetation.[5,11,33,34]

Waves and vegetation can interact according to a range of regimes, which depend on wave and vegetation characteristics.[4,30,35] The wave-vegetation interaction regime determines to what extent bending and swaying of vegetation under wave forces occur, which in turn control to what extent waves are attenuated over vegetation.[36] The vegetation regimes range from rigid vegetation to fully flexible vegetation, and from short waves to long waves.

The engineering challenge lies in efficiently solving the wave damping under the two-way interaction of waves and flexible vegetation. Methods include the calibration of a drag coefficient,[37-40] an effective vegetation length,[41-45] scaling analysis,[4,30,42] cantilever beam models,[11,28,29,36] and complex numerical models that fully solve vegetation motion.[30,34,46] The applicability of these methods depends on the wave-vegetation regime and the available computational resources.

This chapter aims to assist engineers and scientists in solving wave attenuation over flexible salt marsh or seagrass vegetation. We first describe the different components of the wave-vegetation interaction in Section 2. Then, we identify the wave-vegetation interaction regimes and link these to typical field conditions in Section 3. In Section 4, we present a wave damping model that can be applied to the quasi-flexible regime, which is common for coastal vegetation. Finally, the conclusions are presented in Section 5. This chapter focusses on salt marsh and seagrass vegetation, but can also apply to other types of vegetation (e.g. kelp and mangroves) as long as the assumptions used in this chapter are not violated.

2. The wave-vegetation interface

2.1. *Definitions*

We consider monochromatic waves that travel over a field of cylindrical vegetation on top of a flat bed in a direction that is normal to the vegetation field (Figure 1). The cylindrical cross-section represents thin salt marsh grasses. Sea grass blades have rectangular cross-sections.[47] They are dynamically similar, but the different geometry changes the equations. The equations for rectangular-shaped vegetation will not be shown in this chapter, but can be derived in a straightforward manner from the presented analysis. Furthermore, we do not consider tidal currents or wave breaking.

We employ a two-dimensional coordinate system on the scale of the vegetation field. The x-axis is parallel to the direction of wave propagation with $x = 0$ at the seaward edge of the vegetation. The z-axis points upward in vertical direction. The still water surface is located at $z = 0$ and the bed level at $z = -h$, in which h is the still water depth. The waves are defined by their height $H(x)$ and period T. They drive a velocity field $U(x, z, t) = u + iw$, in which the real part u denotes the horizontal velocity component and the complex part w denotes the vertical velocity component. Our two-dimensional coordinate system implies that we consider the wave and vegetation field to be uniform in the horizontal direction perpendicular to the x-axis (y-direction in Figure 1B).

The individual stems in the vegetation are flexible cylinders with height h_v, diameter b_v, mass density ρ_v, and flexural rigidity EI_v. The flexural rigidity $EI_v = \pi b_v^4 E/64$ accounts for both the Young's modulus E_v and the diameter of the vegetation. The density of a vegetation field is expressed as the number of stems per unit area n_v. In case of a staggered vegetation lay-out (1B), the distance between stems equals $S_v = n_v^{-1/2}$. It is assumed that the vegetation stems are inextensible, and have a homogeneous cross-section and flexural rigidity.

The dynamics of the vegetation are expressed on a coordinate system on the scale of individual stems (Figure 1C). The along-stem coordinate s

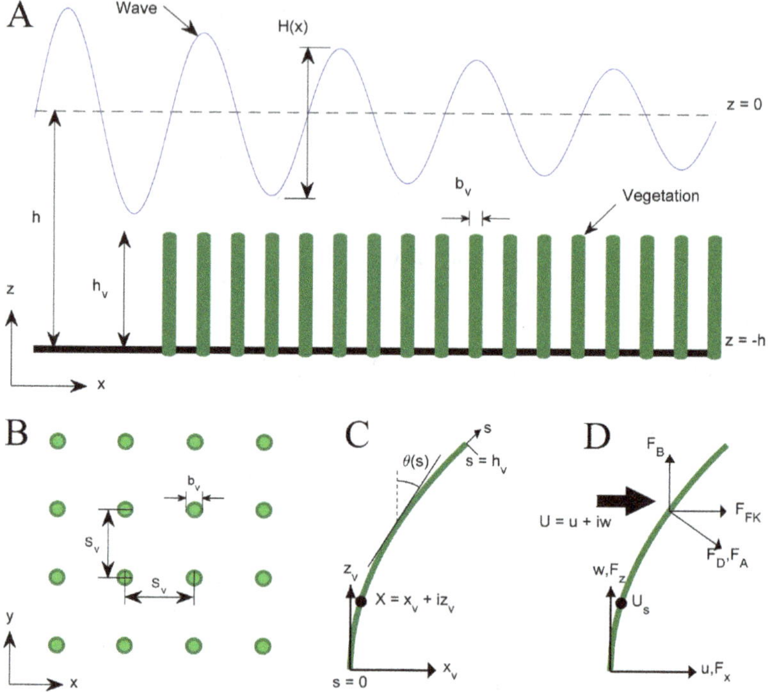

Fig. 1. Definition sketches of the coordinate system of the wave-vegetation interface at the canopy-scale using (A) top and (B) side-view, and at the plant-scale of the (C) stem coordinate system and (D) velocities and forces.

is defined such that $s = 0$ is the root and $s = h_v$ is the tip of the stem. The bending angle $\theta(s,t)$ denotes the curvature in the stem posture. Additionally, cartesian coordinates $x_v(s,t) = \int_0^s \sin\theta$ and $z_v(s,t) = \int_0^s \cos\theta$ are directed in horizontal and vertical direction respectively with their origin at the root of the stem. The posture of the stem is denoted by $X(s,t) = x_v + iz_v$ and the velocity of the stem is defined as $U_{veg} = \partial X / \partial t$. Finally, $U_s(s,t) = u_s + iw_s$ is the wave-driven water particle velocity at the stem, and $F(s,t) = F_x + iF_z$ are the wave forces at the stem (Figure 1D).

2.2. Force balance

The wave-vegetation interface is controlled by wave forces, the buoyancy force, and the internal forces of the vegetation. The wave forces are exerted by the waves on the vegetation. The definitions of the three wave forces[30,32] per unit stem length are: the drag force, which acts in the stem-normal direction;

$$F_D(x, s, t) = \frac{1}{2}\rho C_D b_v |u_{rn}| u_{rn} e^{-i\theta}, \tag{1}$$

the added mass force in the stem-normal direction;

$$F_A(x, s, t) = \frac{1}{4}\pi \rho C_A b_v^2 \frac{\partial u_{rn}}{\partial t} e^{-i\theta}, \tag{2}$$

and the Froude-Krylov force in the flow direction;

$$F_{FK}(x, s, t) = \frac{1}{4}\pi \rho b_v^2 \frac{\partial U_s}{\partial t}. \tag{3}$$

$u_{rn} = \Re\left(U_r e^{i\theta}\right)$ and $u_{rp} = \Im\left(U_r e^{i\theta}\right)$ are the stem-normal and stem-parallel components of the relative velocity between water and stem $U_r = U_s - U_{veg}$. C_D and C_A are coefficients for the drag and the added mass force respectively, and ρ is the mass density of water. The contribution of skin friction is omitted, due to the small values of the friction coefficient under field wave conditions.[29,32]

In case of rigid vegetation, the stems do not move, i.e. $U_{veg} = 0$ and $U_r = U_s$. Furthermore, rigid stems remain upright such that the stem-normal direction is always parallel to the x-axis and the stem-parallel direction is always parallel to the z-axis, whereas the stem-normal and parallel directions of flexible stems change over a wave cycle following the time-dependent bending angle of the stems.

The stem buoyancy force results from the density difference between vegetation and water. It acts in a direction parallel to the z-axis. The buoyancy force of cylindrical stems per unit stem length is given by[30]

$$F_B(s, t) = i\frac{1}{4}\pi(\rho - \rho_v)g b_v^2 \tag{4}$$

The vegetation forces are the stem inertia and the mechanical forces that resist stem deformation. The stem inertia force of a cylindrical stem per unit stem length is defined as

$$F_I(x, s, t) = \frac{1}{4}\pi \rho_s b_v^2 \frac{\partial U_{veg}}{\partial t}. \tag{5}$$

The two mechanical forces are the shear and tension forces. The shear stress per unit stem length is defined as

$$F_S(x, s, t) = \frac{\partial}{\partial s}EI_v \frac{\partial^2 \theta}{\partial s^2} e^{-i\theta} \tag{6}$$

and resists stem-normal deformation. The tension force

$$F_T(x, s, t) = -\frac{\partial}{\partial s}\hat{T}i e^{-i\theta} \tag{7}$$

with tension $\hat{T}$ acts in stem-parallel direction. The tension force closes the stem-parallel force balance such that the inextensibility condition is satisfied.

The wave forces, the buoyancy force, and the vegetation forces collectively form the force balance which controls the wave-vegetation interaction. The force balance satisfies

$$F_I + F_S + F_T = F_D + F_A + F_{FK} + F_B. \tag{8}$$

The vegetation forces on the left-hand side balance the wave and buoyancy forces.

2.3. *Scaling analysis*

We conduct a scaling of our system in order to identify the relative magnitude of the terms in the force balance (Eq. 8). The normalized terms are identified by an asterisk in their subscript. We introduce scaled coordinates

$$x_* = \frac{x}{A_w}, \quad z_* = \frac{z}{h_v}, \quad s_* = \frac{s}{h_v}, \quad t_* = t\omega \tag{9}$$

and quantities

$$X_* = \frac{X}{h_v}, \qquad H_* = \frac{H}{h}, \qquad [U_*, U_{r*}] = \frac{[U, U_r]}{u_c}, \qquad U_{veg*} = \frac{U_{veg}}{h_v\omega},$$

$$F_* = \frac{F}{\rho b_v u_c^2}, \qquad [F_{S*}, F_{T*}] = \frac{[F_S, F_T]\, h_v^3}{EI_v}, \qquad \hat{T}_* = \frac{\hat{T} h_v^2}{EI_v}. \tag{10}$$

Herein, $\omega = 2\pi/T$ is the wave angular frequency, u_c is the velocity scale, and $A_w = u_c/\omega$ is the typical wave excursion length. The mechanical forces F_S and F_T have been scaled with EI_v/h_v^3 rather than velocity as they depend on mechanical properties instead of velocity.

When Eq. (1)-(7) are normalized and substituted into Eq. (8), the force balance can be written as.[30]

$$\underbrace{\frac{1}{2}\frac{\pi^2}{KC}\rho'CaL\frac{\partial U_{veg*}}{\partial t_*}}_{\text{inertia}} + \left(\underbrace{\frac{\partial^3\theta}{\partial s_*^3} - i\frac{\partial\theta}{\partial s_*}\frac{\partial^2\theta}{\partial s_*^2}}_{\text{shear}} + \underbrace{\hat{T}_*\frac{\partial\theta}{\partial s_*} + i\frac{\partial\hat{T}_*}{\partial s_*}}_{\text{tension}} \right)e^{-i\theta} =$$

$$\frac{1}{2}Ca\left(\underbrace{C_D|u_{rn*}|u_{rn*}}_{\text{drag}} + \underbrace{C_A\frac{\pi^2}{KC}\frac{\partial u_{rn*}}{\partial t_*}}_{\text{added mass}} + \underbrace{\frac{\pi^2}{KC}\frac{\partial U_{s*}}{\partial t_*}}_{\text{Froude–Krylov}} \right) + \underbrace{iB}_{\text{buoyancy}} \tag{11}$$

with $\rho' = \rho_v/\rho$. The real part of Eq. (11) denotes the force balance in horizontal (x_*) direction and the complex part describes the force balance in vertical (z_*) direction.

The scaling analysis has revealed four dimensionless numbers that control the wave vegetation interaction. First, the Cauchy number[48]

$$Ca = \frac{\rho b_v u_c^2 h_v^3}{EI_v} \tag{12}$$

represents the ratio between wave forces and bending resistance. $Ca \ll 1$ corresponds with rigid vegetation – the stiffness is much larger than the wave forces – and $Ca \gg 1$ corresponds with flexible vegetation – the stiffness is much smaller than the wave forces. Second, the excursion ratio[30]

$$L = \frac{h_v}{A_w} \tag{13}$$

is the ratio between stem length and wave excursion length. Third, the Buoyancy number[48]

$$B = \frac{(\rho_v - \rho)\,\pi g b_v^2 h_v^3}{4EI_v} \tag{14}$$

represents the ratio between the buoyancy force and bending resistance. $L \ll 1$ corresponds to long waves with a wave excursion length much larger than the height of the vegetation. Conversely, $L \gg 1$ corresponds to short waves with excursion length smaller than the height of the vegetation. Fourth, The Keulegan-Carpenter number[49]

$$KC = \frac{u_c T}{b_v} \tag{15}$$

is the ratio between wave particle excursion and stem diameter.

Indicative values for these dimensionless ratios in coastal vegetation canopies are presented in Table 1. The ratios depend on both hydrodynamic and vegetation conditions with the exception of B which is function of vegetation characteristics only. Therefore, the ratio of these dimensionless numbers may differ strongly between locations and wave events. The values of Ca and KC are typically higher for seagrass than for salt marsh vegetation, because seagrass is often thinner than salt marsh vegetation.

3. Wave-vegetation interaction regimes

We can identify different wave-vegetation interaction regimes based on the scaling analysis. In most practical cases, the buoyancy number is small as

Table 1. Typical values for the Cauchy number Ca, excursion ratio L, buoyancy number B, and the Keulegan-Carpenter number KC of salt marsh and seagrass vegetation. The rigid values of Ca and B are provided for reference.

Parameter	Salt marsh[14,17,22,24,29,50–56]	Seagrass*[10,13,43,47,57–60]	Rigid
Ca	10^{-1} - 10^3	10^0 - 10^5	0
L	10^{-1} - 10^2	10^{-1} - 10^2	–
B	0 - $10^{1\dagger}$	0 - 10^2	0
KC	10^1 - 10^4	10^2-10^5	–

*Based on Lei and Nepf[43] and references therein.
†Estimated using $\rho_v = 700$ kg/m^3.

the density of the coastal vegetation is similar to the density of water[43] (Table 1). Then, the wave-vegetation interaction regimes depend on Ca and L. We can identify five different regimes, as presented in the regime diagram (Figure 2).

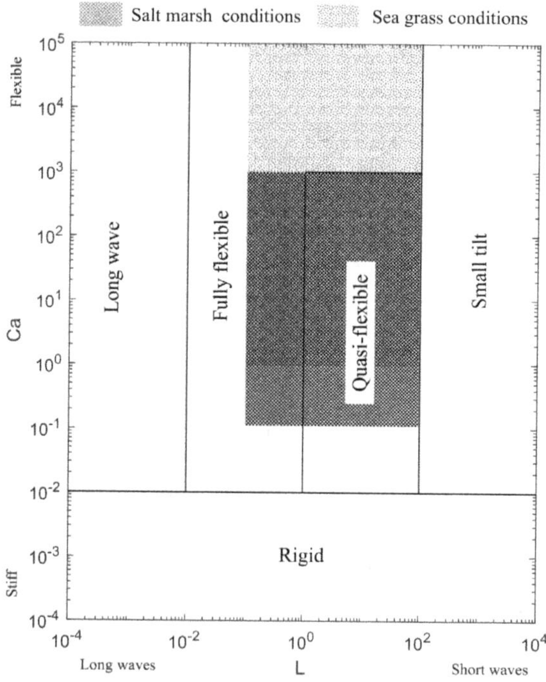

Fig. 2. Regime diagram of wave-vegetation conditions as function of the Cauchy number Ca and the excursion ratio L. The shaded areas denote typical conditions for salt marsh and seagrass habitats under coastal wave conditions.

Rigid The rigid vegetation regime occurs when $Ca \ll 1$. The wave forces are much smaller than the bending resistance of a stem. Any bending of the vegetation will be negligible. The wave damping over vegetation satisfies rigid vegetation models.[61]

Small tilt The small tilt regime occurs when $L \gg 1$. The wave excursion length is much shorter than the stem length. The vegetation stems will follow the wave motion, but their excursion over a wave cycle is limited by the short waves. Therefore, the vegetation motion is primarily in the form of horizontal motion with a small bending angle. The horizontal motion reduces the relative velocity, which reduces wave damping.[28] The wave-induced horizontal motion of the vegetation controls the wave damping.

Quasi-flexible The quasi-flexible vegetation regime occurs when $L > 1$ and $Ca < 10^3$. The wave excursion length is equal to or shorter than the vegetation length. The vegetation moves significantly over a wave cycle with its maximum extension up to the wave excursion length. The vegetation exhibits motion in horizontal and vertical directions and its posture may feature smooth to sharp bends. The motion and bending are stronger in the top section of the stem than in the bottom section near the root. Flume experiments[29] have shown that wave damping by quasi-flexible vegetation is primarily driven by the upright bottom section of a stem.

Fully flexible The fully flexible vegetation regime occurs when $L < 1$ or when $L > 1$ and $Ca > 10^3$. The wave excursion length may be longer than the length of the vegetation. The vegetation may fully extend during part of the wave cycle, in which case the vegetation motion may be restricted by its anchor at the root. The vegetation may also exhibit extreme bending when Ca is very large. The motion of vegetation, the absence of motion when the stem is fully extended, and the bending angles affect the wave damping by fully flexible vegetation. The motion and the lack of it when fully extended affect the relative velocity over a wave cycle. Furthermore, the sharp bending angles affect the magnitude and the angle of wave forces. Both the relative velocity and bending effects control the wave damping by fully flexible vegetation.[30,46]

Long wave The long wave regime extends towards unidirectional flow for very long waves (e.g. tides). It occurs when $L \ll 1$. The wave excursion length is much longer than the stem length. The rate of change of flow velocity over time is typically low (near steady).

The vegetation can bend sharply, but at low vegetation velocities. Therefore, wave damping is this regime is primarily driven by the bending angle of the vegetation, which changes force direction and magnitude.

The transitions between adjacent regimes are gradual. The boundaries in the regime diagram (Figure 2) indicate the orders of magnitude around which the transitions occur. The precise locations of the boundaries are subject to ongoing research.

When the buoyancy number is not small and $B > \sqrt{CaL}$, the wave-vegetation regime is dominated by buoyancy.[36] Wave damping is controlled by the balance between wave forcing and the restoring force from buoyancy. This regime has been modelled by Henderson.[36]

4. Modelling wave attenuation by quasi-flexible vegetation

The quasi-flexible vegetation regime is common for salt marsh vegetation and seagrass under conditions with wind waves, as is revealed by overlaying the indicative values from Table 1 with the regime diagram (Figure 2). Furthermore, solving wave damping in the quasi-flexible regime is computationally cheaper than the fully flexible regime.[29] Therefore, the wave damping by quasi-flexible vegetation is covered here. The framework presented here is also applicable to the small tilt and rigid regimes. When the conditions fall within the fully flexible regime, complex numerical models are currently required to solve the vegetation motion.[30,34] Finally, steady flow models[48,62,63] can be applied to study the long wave regime.

4.1. *Model assumptions*

Experiments[29] with a range of quasi-flexible vegetation conditions have revealed that: (i) the wave energy dissipation is dominated by the work done by the drag force; (ii) the dissipation is concentrated at the upright bottom section of a flexible vegetation stem; (iii) the impact of vegetation motion on the relative velocity is more important for wave attenuation than the effect of a bent plant posture on force direction and magnitude.

These observations justify the following model assumptions. It is assumed that

(1) wave energy is dissipated where plant deflections are small, and the plant posture is near-vertical);

(2) the drag force controls the wave-vegetation interaction;
(3) vegetation is thin, cylindrical, inextensible, with homogeneous cross-sections and flexural rigidity (following Section 2.1);
(4) stem-stem interactions can be neglected.

Under these assumption, the wave-vegetation interaction is represented by near-vertical stems that can move in x-direction and are constrained by the root. This model formulation effectively reduces the quasi-flexible regime to that of the small tilt regime as studied in Mullarney and Henderson.[28]

4.2. *Simplified force balance*

The force balance (Eq. 11) can be simplified under the model assumptions as a stem-normal force balance (x-direction only). The drag force is the only wave force included. The stem-parallel contributions of buoyancy and tension can be omitted. The contribution of stem inertia is negligible for thin stems ($\pi^2 \rho'/(2KC) \ll 1$). Furthermore, x_v is now normalized by horizontal water particle excursion length A_w instead of h_v to reflect that the stem is near-vertical. The bending angle is approximated by $\theta \approx \partial x_{v*}/\partial z_{v*}$, and the along-stem coordinate by $s_* \approx z_{v*}$ for near-vertical stems.[28,30] Under these conditions, Eq. (11) simplifies to

$$\frac{\partial^4 x_{v*}}{\partial z_{v*}^4} = Q\left(u_{s*} - \frac{\partial x_{v*}}{\partial t_*}\right), \tag{16}$$

in which scaled flexibility

$$Q = \frac{4}{3\pi}C_D CaL \int_0^1 (a_u - a_v)dz_{v*} \tag{17}$$

is a linearised parameterisation of the magnitude of drag force, and a_u and a_v are the amplitudes of the water velocity at the stem and the vegetation velocity respectively. The boundary conditions of Eq. (16) are defined as clamped at the root, $x_{v*} = \partial x_{v*}/\partial z_{v*} = 0$ at $z_{v*} = 0$, and free at the tip, $\partial^2 x_{v*}/\partial z_{v*}^2 = \partial^3 x_{v*}/\partial z_{v*}^3 = 0$ at $z_{v*} = 1$.

Eq. (16) is structured as a cantilever beam problem. The stem motion as function of the wave and vegetation condition can be obtained via a transfer function.[28] The transfer function of a monochromatic wave follows from expanding the problem parameters into spatiotemporal modes with a single temporal mode and N spatial modes, such that $u_{s*} = a_u \Re\left(e^{i(t_*+\phi_u)}\right) = \Re\left(e^{it_*}\sum_{n=1}^{N} U_n \psi_n\right)$ with spatiotemporal coefficient U_n and spatial modes

ψ_n. The spatial modes satisfy $\partial \psi_n / \partial z_{v*} = \alpha_n \psi_n$ where α_n are the eigenvalues of each spatial mode.

It follows that the transfer function between the water particle velocity at the stem velocity and the relative velocity is given by

$$R(z_{v*}) = 1 - \frac{\sum_{n=1}^{N} \left(\frac{1}{1 - \frac{i \alpha_n^4}{Q}} U_n \psi_n \right)}{\sum_{n=1}^{N} U_n \psi_n}. \tag{18}$$

The transfer function R is a complex number with an amplitude and a phase according to $R = a_R e^{i\phi_R}$. a_R denotes the amplitude transfer from u_{s*} to u_{r*}. a_R equals 1 for rigid vegetation and reduces as vegetation motion increasingly follows wave motion. ϕ_R denotes the phase difference between u_{s*} and u_{r*}. As Q is a function of the vegetation velocity amplitude a_v, Eq. (18) must be solved iteratively. It can be shown that the solution is unique.[29]

4.3. *Solution of wave damping by flexible vegetation*

The transfer function can be used to define a depth-dependent work factor χ that represents the reduction of the work done by a wave over a flexible stem relative to the work that would be done over a rigid stem.

$$\chi(z_{v*}) = \frac{\overline{W}_*}{\overline{W}_{rig*}}. \tag{19}$$

$\overline{W}_* = \overline{F_{D*} u_{s*}}$ is the wave-averaged work done over a flexible stem and $\overline{W}_{rig*}$ is the work done over a rigid stem with identical geometry. By substituting the normalized equivalent of Eq. (1), $u_{s*} = a_u \Re \left(e^{i(t_* + \phi_u)} \right)$, and $u_{r*} = a_R a_U \Re \left(e^{i(t_* + \phi_u + \phi_R)} \right)$ into Eq. (19), it follows that

$$\chi = a_R^2 \cos \phi_R. \tag{20}$$

The work factor is a function the scaled flexibility and the along-stem coordinate (Figure 3). It decreases when the wave-vegetation conditions become more flexible (increasing Q). Furthermore, the work factor is higher at the root of the vegetation than at the tip, which is consistent with the model assumption that the wave damping is controlled by the lower section of the stem (Section 4.1). Under specific wave-vegetation conditions, the vegetation velocity may locally exceed the water velocity,[28] which leads to a negative contribution to wave damping.[29]

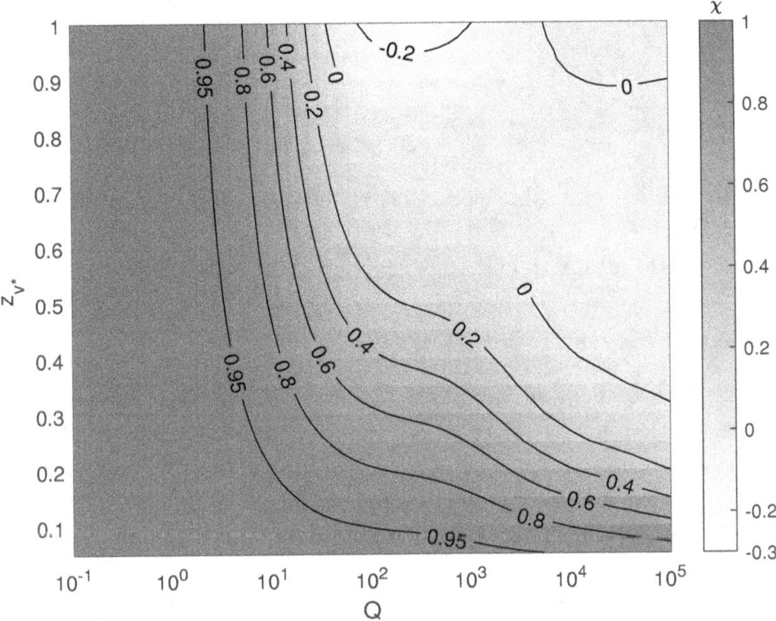

Fig. 3. Contour plot of the work factor χ as a function of the scaled flexibility Q and normalized along-stem coordinate z_v* under a velocity profile that satisfies linear wave theory with $kh_v = 0.5$. The scaled flexibility Q on the x-axis is at the first iteration, which uses $a_v = 0$. The number of spatial modes used in the solution procedure is $N = 20$.

The work factor χ can be applied in different wave damping models that use a range of assumptions in addition to those stated in Section 4.1. Wave damping can be expressed via a damping coefficient $\beta(x)$ that satisfies[61] (not normalized)

$$H(x) = \frac{H_0}{1 + \beta x},\qquad(21)$$

in which H_0 is the wave height at the leading edge of the vegetation field, i.e. at $x = 0$. β is spatially uniform when the vegetation is rigid, and a function of x when the vegetation is flexible. The wave-vegetation interaction changes as the waves are attenuated over a vegetation field. As the wave height decreases, Q decreases such that the wave-vegetation interaction moves closer to that of rigid vegetation.

When linear wave theory is valid, the wave damping coefficient can be expressed as

$$\beta = \frac{4}{3\pi} C_D n_v b_v H_0 k^2 \frac{\int_{z_v=0}^{h_v} \chi \cosh^3 (kz_v)\, dz_v}{(\sinh 2kh + 2kh) \sinh kh} \tag{22}$$

with k being the wave number. Eq. (22) reduces to the formulation of β for rigid vegetation[33,61] when $\chi = 1$. Eq. (22) has been successfully validated for a range of real and artificial coastal vegetation types.[29]

4.4. *Link to effective vegetation length*

A representation of flexible vegetation in wave damping models is via an effective vegetation length h_e.[41–45] h_e is defined as the length of a rigid vegetation stem that contributes equally to wave damping as a flexible vegetation stem of length h_v, i.e. $\beta(x, h_{e,rig}(x)) = \beta(x, h_{v,flex}(x))$. The effective vegetation length can be computed from χ by solving

$$\int_{z_v=0}^{h_v} \chi \cosh^3 (kz_v)\, dz_v = \frac{1}{3k} \left(\sinh^3 (kh_e) + 3\sinh (kh_e) \right) \tag{23}$$

when linear wave theory is assumed. The work factor and the effective vegetation length are conceptually similar, as they both represent a reduction of the wave damping of flexible vegetation relative to rigid vegetation.[31] The effective vegetation length typically relies on a calibration.[31,41–43] The work factor does not and may thus provide a robust basis when no calibration data is available. Furthermore, the work factor allows for a greater variation in the wave velocity profile, including vegetation-induced amplifications and depressions in the velocity profile,[24] as it does not modify the vegetation length.

5. Conclusions

The wave-vegetation interaction can be structured into five non-buoyant regimes and one buoyant regime. The five buoyant regimes can be classified according to the Cauchy number and the excursion ratio. The wave-vegetation regime controls the relative impact of bending and swaying on wave damping. Plant bending is dominant in the long wave regime, whereas plant swaying is dominant in the small tilt and quasi-flexible regimes. Both bending and swaying are important in the fully flexible regime, and neither occurs in the rigid regime. The quasi-flexible regime is the most common for salt marsh grasses and seagrass blades under wind wave conditions, although the other regimes may also occur.

Wave attenuation in the quasi-flexible regime depends on the scaled flexibility Q, which can be computed from the wave conditions and vegetation characteristics. The reduction of wave damping by flexible vegetation compared to rigid counterparts of identical geometry is expressed as function of work factor χ and is a function of the scaled flexibility. The work factor can be implemented in wave damping models to account for vegetation flexibility. For example, the effective vegetation length can be derived from the work factor.

The wave damping model presented here is subject to limitations. The model is valid for the quasi-flexible, small tilt and rigid regimes, but not for the fully flexible and long wave regimes. These regimes require complex numerical models and unidirectional flow models respectively. We have also restricted ourselves conditions with limited buoyancy, monochromatic waves, and cylindrical stems. While the conceptual analysis remains similar, the model will requires specific modifications to overcome these restrictions. Finally, the model may be also applicable to other coastal vegetation species, such as kelp and mangroves, in addition to salt marsh and seagrass canopies when they satisfy the conditions considered here.

References

1. J. C. Mullarney and S. M. Henderson. Flows Within Marine Vegetation Canopies. In eds. V. Panchang and J. Kaihatu, *Advances in Coastal Hydraulics*, pp. 1–46. World Scientific (2018). ISBN 978-981-323-127-6. doi: 10.1142/9789813231283_0001.

2. J. A. Tempest, I. Möller, and T. Spencer, A Review of Plant-Flow Interactions on Salt Marshes: The Importance of Vegetation Structure and Plant Mechanical Characteristics, *WIREs Water.* **2**(6), 669–681 (Nov., 2015). ISSN 2049-1948. doi: 10.1002/wat2.1103.

3. F. Rupprecht, I. Möller, M. Paul, M. Kudella, T. Spencer, B. K. van Wesenbeeck, G. Wolters, K. Jensen, T. J. Bouma, M. Miranda-Lange, and S. Schimmels, Vegetation-Wave Interactions in Salt Marshes under Storm Surge Conditions, *Ecological Engineering.* **100**, 301–315 (Mar., 2017). ISSN 0925-8574. doi: 10.1016/j.ecoleng.2016.12.030.

4. T. Leclercq and E. de Langre, Reconfiguration of elastic blades in oscillatory flow, *Journal of Fluid Mechanics.* **838**, 606–630 (Mar., 2018). ISSN 0022-1120, 1469-7645. doi: 10.1017/jfm.2017.910.

5. N. G. Jacobsen, W. Bakker, W. S. J. Uijttewaal, and R. Uittenbogaard,

Experimental investigation of the wave-induced motion of and force distribution along a flexible stem, *Journal of Fluid Mechanics.* **880**, 1036–1069 (Dec., 2019). ISSN 0022-1120, 1469-7645. doi: 10.1017/jfm.2019.739.

6. S. Temmerman, P. Meire, T. J. Bouma, P. M. J. Herman, T. Ysebaert, and H. J. De Vriend, Ecosystem-based coastal defence in the face of global change, *Nature.* **504**(7478), 79–83 (Dec., 2013). ISSN 1476-4687. doi: 10.1038/nature12859.

7. N. Leonardi, I. Carnacina, C. Donatelli, N. K. Ganju, A. Plater, M. Schuerch, and S. Temmerman, Dynamic interactions between coastal storms and salt marshes: A review, *Geomorphology.* **301**, 92–107 (Jan., 2018). doi: 10.1016/j.geomorph.2017.11.001.

8. B. W. Borsje, B. K. van Wesenbeeck, F. Dekker, P. Paalvast, T. J. Bouma, M. M. van Katwijk, and M. B. de Vries, How Ecological Engineering Can Serve in Coastal Protection, *Ecological Engineering.* **37**(2), 113–122 (Feb., 2011). ISSN 0925-8574. doi: 10.1016/j.ecoleng.2010.11.027.

9. M. L. Pinsky, G. Guannel, and K. K. Arkema, Quantifying wave attenuation to inform coastal habitat conservation, *Ecosphere.* **4**(8), art95 (2013). ISSN 2150-8925. doi: 10.1890/ES13-00080.1.

10. K. Bradley and C. Houser, Relative velocity of seagrass blades: Implications for wave attenuation in low-energy environments, *Journal of Geophysical Research: Earth Surface.* **114**(F1) (2009). ISSN 2156-2202. doi: 10.1029/2007JF000951.

11. L. Zhu, K. Huguenard, Q.-P. Zou, D. W. Fredriksson, and D. Xie, Aquaculture farms as nature-based coastal protection: Random wave attenuation by suspended and submerged canopies, *Coastal Engineering.* **160**, 103737 (Sept., 2020). ISSN 0378-3839. doi: 10.1016/j.coastaleng.2020.103737.

12. M. Mork, The effect of kelp in wave damping, *Sarsia.* **80**(4), 323–327 (Feb., 1996). ISSN 0036-4827. doi: 10.1080/00364827.1996.10413607.

13. E. Infantes, A. Orfila, G. Simarro, J. Terrados, M. Luhar, and H. Nepf, Effect of a seagrass (Posidonia oceanica) meadow on wave propagation, *Marine Ecology Progress Series.* **456**, 63–72 (June, 2012). ISSN 0171-8630, 1616-1599. doi: 10.3354/meps09754.

14. R. S. Jadhav, Q. Chen, and J. M. Smith, Spectral distribution of wave energy dissipation by salt marsh vegetation, *Coastal Engineering.* **77**, 99–107 (July, 2013). ISSN 0378-3839. doi: 10.1016/j.coastaleng.2013.02.013.

15. J. L. Garzon, M. Maza, C. M. Ferreira, J. L. Lara, and I. J. Losada, Wave Attenuation by Spartina Saltmarshes in the Chesapeake Bay Under Storm Surge Conditions, *Journal of Geophysical Research: Oceans.* **124**(7), 5220–5243 (2019). ISSN 2169-9291. doi: 10.1029/2018JC014865.

16. A. Dubi and A. Tørum, Wave Damping by Kelp Vegetation, *Coastal Engineering 1994* (1995). ISSN 9780784400890. doi: 10.1061/9780784400890.012.

17. I. Möller, M. Kudella, F. Rupprecht, T. Spencer, M. Paul, B. K. van Wesen-

beeck, G. Wolters, K. Jensen, T. J. Bouma, M. Miranda-Lange, and S. Schimmels, Wave Attenuation over Coastal Salt Marshes under Storm Surge Conditions, *Nature Geoscience.* **7**(10), 727–731 (Oct., 2014). ISSN 1752-0894. doi: 10.1038/ngeo2251.

18. M. Maza, J. L. Lara, I. J. Losada, B. Ondiviela, J. Trinogga, and T. J. Bouma, Large-Scale 3-D Experiments of Wave and Current Interaction with Real Vegetation. Part 2: Experimental Analysis, *Coastal Engineering.* **106**, 73–86 (Dec., 2015). ISSN 0378-3839. doi: 10.1016/j.coastaleng.2015.09.010.

19. J. F. Sánchez-González, V. Sánchez-Rojas, and C. D. Memos, Wave attenuation due to Posidonia oceanica meadows, *Journal of Hydraulic Research.* **49**(4), 503–514 (2011). doi: 10.1080/00221686.2011.552464.

20. L. Zhu, J. Lei, K. Huguenard, and D. W. Fredriksson, Wave attenuation by suspended canopies with cultivated kelp (Saccharina latissima), *Coastal Engineering.* **168**, 103947 (Sept., 2021). ISSN 0378-3839. doi: 10.1016/j. coastaleng.2021.103947.

21. V. Vuik, B. W. Borsje, P. W. J. M. Willemsen, and S. N. Jonkman, Salt marshes for flood risk reduction: Quantifying long-term effectiveness and life-cycle costs, *Ocean & Coastal Management.* **171**, 96–110 (Apr., 2019). ISSN 0964-5691. doi: 10.1016/j.ocecoaman.2019.01.010.

22. W. G. Bennett, T. J. van Veelen, T. P. Fairchild, J. N. Griffin, and H. Karunarathna, Computational Modelling of the Impacts of Saltmarsh Management Interventions on Hydrodynamics of a Small Macro-Tidal Estuary, *Journal of Marine Science and Engineering.* **8**(5), 373 (May, 2020). doi: 10.3390/jmse8050373.

23. T. P. Fairchild, W. G. Bennett, G. Smith, B. Day, M. W. Skov, I. Möller, N. Beaumont, H. Karunarathna, and J. N. Griffin, Coastal wetlands mitigate storm flooding and associated costs in estuaries, *Environmental Research Letters.* **16**(7), 074034 (July, 2021). ISSN 1748-9326. doi: 10.1088/1748-9326/ ac0c45.

24. T. J. van Veelen, T. P. Fairchild, D. E. Reeve, and H. Karunarathna, Experimental study on vegetation flexibility as control parameter for wave damping and velocity structure, *Coastal Engineering.* **157**, 103648 (2020).

25. M. Paul, F. Rupprecht, I. Möller, T. J. Bouma, T. Spencer, M. Kudella, G. Wolters, B. K. van Wesenbeeck, K. Jensen, M. Miranda-Lange, and S. Schimmels, Plant Stiffness and Biomass as Drivers for Drag Forces under Extreme Wave Loading: A Flume Study on Mimics, *Coastal Engineering.* **117**, 70–78 (Nov., 2016). ISSN 0378-3839. doi: 10.1016/j.coastaleng.2016.07. 004.

26. K. C. Riffe, S. M. Henderson, and J. C. Mullarney, Wave dissipation by flexible vegetation, *Geophys Research Letters.* **38**(18), L18607 (2011). doi: 10.1029/2011GL048773.

27. F. J. Méndez, I. J. Losada, and M. A. Losada, Hydrodynamics induced

by wind waves in a vegetation field, *Journal of Geophysical Research: Oceans.* **104**(C8), 18383–18396 (Aug., 1999). ISSN 2156-2202. doi: 10.1029/ 1999JC900119.

28. J. C. Mullarney and S. M. Henderson, Wave-Forced Motion of Submerged Single-Stem Vegetation, *Journal of Geophysical Research: Oceans.* **115**(C12) (Dec., 2010). ISSN 2156-2202. doi: 10.1029/2010JC006448.

29. T. J. van Veelen, H. Karunarathna, and D. E. Reeve, Modelling wave attenuation by quasi-flexible coastal vegetation, *Coastal Engineering.* **164**, 103820 (Mar., 2021). ISSN 0378-3839. doi: 10.1016/j.coastaleng.2020.103820.

30. M. Luhar and H. M. Nepf, Wave-Induced Dynamics of Flexible Blades, *Journal of Fluids and Structures.* **61**, 20–41 (Feb., 2016). ISSN 0889-9746. doi: 10.1016/j.jfluidstructs.2015.11.007.

31. L. Zhu, K. Huguenard, D. W. Fredriksson, and J. Lei, Wave attenuation by flexible vegetation (and suspended kelp) with blade motion: Analytical solutions, *Advances in Water Resources.* **162**, 104148 (Apr., 2022). ISSN 0309-1708. doi: 10.1016/j.advwatres.2022.104148.

32. R. B. Zeller, J. S. Weitzman, M. E. Abbett, F. J. Zarama, O. B. Fringer, and J. R. Koseff, Improved parameterization of seagrass blade dynamics and wave attenuation based on numerical and laboratory experiments, *Limnology and Oceanography.* **59**(1), 251–266 (2014). ISSN 1939-5590. doi: 10.4319/lo. 2014.59.1.0251.

33. F. J. Mendez and I. J. Losada, An Empirical Model to Estimate the Propagation of Random Breaking and Nonbreaking Waves over Vegetation Fields, *Coastal Engineering.* **51**(2), 103–118 (Apr., 2004). ISSN 0378-3839. doi: 10.1016/j.coastaleng.2003.11.003.

34. H. Chen and Q.-P. Zou, Eulerian-Lagrangian flow-vegetation interaction model using immersed boundary method and OpenFOAM, *Advances in Water Resources.* **126**, 176–192 (Feb., 2019). ISSN 0309-1708. doi: 10.1016/j. advwatres.2019.02.006.

35. R. J. Lowe, J. R. Koseff, and S. G. Monismith, Oscillatory Flow through Submerged Canopies: 1. Velocity Structure, *Journal of Geophysical Research: Oceans.* **110**(C10) (Oct., 2005). ISSN 0148-0227. doi: 10.1029/ 2004JC002788.

36. S. M. Henderson, Motion of buoyant, flexible aquatic vegetation under waves: Simple theoretical models and parameterization of wave dissipation, *Coastal Engineering.* **152**, 103497 (Oct., 2019). ISSN 0378-3839. doi: 10.1016/j.coastaleng.2019.04.009.

37. N. Kobayashi, A. W. Raichle, and T. Asano, Wave Attenuation by Vegetation, *Journal of Waterway, Port, Coastal, and Ocean Engineering.* **119**(1), 30–48 (1993). doi: 10.1061/(ASCE)0733-950X(1993)119:1(30).

38. Y. Ozeren, D. G. Wren, and W. Wu, Experimental Investigation of Wave Attenuation through Model and Live Vegetation, *Journal of Waterway, Port,*

Coastal, and Ocean Engineering. **140**(5), 04014019 (Sept., 2014). doi: 10. 1061/(ASCE)WW.1943-5460.0000251.

39. V. Vuik, S. N. Jonkman, B. W. Borsje, and T. Suzuki, Nature-Based Flood Protection: The Efficiency of Vegetated Foreshores for Reducing Wave Loads on Coastal Dikes, *Coastal Engineering.* **116**, 42–56 (Oct., 2016). ISSN 0378-3839. doi: 10.1016/j.coastaleng.2016.06.001.

40. H. Chen, Y. Ni, Y. Li, F. Liu, S. Ou, M. Su, Y. Peng, Z. Hu, W. Uijt-tewaal, and T. Suzuki, Deriving vegetation drag coefficients in combined wave-current flows by calibration and direct measurement methods, *Advances in Water Resources.* **122**, 217–227 (Dec., 2018). ISSN 0309-1708. doi: 10.1016/j.advwatres.2018.10.008.

41. I. J. Losada, M. Maza, and J. L. Lara, A New Formulation for Vegetation-Induced Damping under Combined Waves and Currents, *Coastal Engineering.* **107**, 1–13 (Jan., 2016). ISSN 0378-3839. doi: 10.1016/j.coastaleng.2015. 09.011.

42. M. Luhar, E. Infantes, and H. Nepf, Seagrass blade motion under waves and its impact on wave decay, *Journal of Geophysical Research: Oceans.* **122**(5), 3736–3752 (2017). ISSN 2169-9291. doi: 10.1002/2017JC012731.

43. J. Lei and H. Nepf, Wave damping by flexible vegetation: Connecting individual blade dynamics to the meadow scale, *Coastal Engineering.* **147**, 138–148 (May, 2019). ISSN 0378-3839. doi: 10.1016/j.coastaleng.2019.01.008.

44. X. Zhang and H. Nepf, Wave-induced reconfiguration of and drag on marsh plants, *Journal of Fluids and Structures.* **100**, 103192 (Jan., 2021). ISSN 0889-9746. doi: 10.1016/j.jfluidstructs.2020.103192.

45. Y. Ding, Q. J. Chen, L. Zhu, J. D. Rosati, and B. D. Johnson. Implementation of flexible vegetation into CSHORE for modeling wave attenuation. Report, Engineer Research and Development Center (U.S.) (Feb., 2022).

46. M. Maza, J. L. Lara, and I. J. Losada, A Coupled Model of Submerged Vegetation under Oscillatory Flow Using Navier–Stokes Equations, *Coastal Engineering.* **80**, 16–34 (Oct., 2013). ISSN 0378-3839. doi: 10.1016/j.coastaleng. 2013.04.009.

47. M. A. Abdelrhman, Modeling coupling between eelgrass Zostera marina and water flow, *Marine Ecology Progress Series.* **338**, 81–96 (May, 2007). ISSN 0171-8630, 1616-1599. doi: 10.3354/meps338081.

48. M. Luhar and H. M. Nepf, Flow-induced Reconfiguration of Buoyant and Flexible Aquatic Vegetation, *Limnology and Oceanography.* **56**(6), 2003–2017 (Sept., 2011). ISSN 0024-3590. doi: 10.4319/lo.2011.56.6.2003.

49. G. H. Keulegan and L. H. Carpenter, Forces on cylinders and plates in an oscillating fluid, *Journal of Research of the National Bureau of Standards Research Paper.* **2857**, 423–440 (1958).

50. J. L. Lara, M. Maza, B. Ondiviela, J. Trinogga, I. J. Losada, T. J. Bouma, and N. Gordejuela, Large-Scale 3-D Experiments of Wave and Current In-

teraction with Real Vegetation. Part 1: Guidelines for Physical Modeling, *Coastal Engineering.* **107**, 70–83 (Jan., 2016). ISSN 0378-3839. doi: 10.1016/j.coastaleng.2015.09.012.

51. J. Chatagnier, The Biomechanics of Salt Marsh Vegetation Applied to Wave and Surge Attenuation, *LSU Masters Theses* (Jan., 2012).

52. R. A. Feagin, J. L. Irish, I. Möller, A. M. Williams, R. J. Colón-Rivera, and M. E. Mousavi, Short Communication: Engineering Properties of Wetland Plants with Application to Wave Attenuation, *Coastal Engineering.* **58**(3), 251–255 (Mar., 2011). ISSN 0378-3839. doi: 10.1016/j.coastaleng.2010.10.003.

53. D. Schulze, F. Rupprecht, S. Nolte, and K. Jensen, Seasonal and spatial within-marsh differences of biophysical plant properties: Implications for wave attenuation capacity of salt marshes, *Aquat Sci.* **81**(4), 65 (July, 2019). ISSN 1420-9055. doi: 10.1007/s00027-019-0660-1.

54. F. Rupprecht, I. Möller, B. Evans, T. Spencer, and K. Jensen, Biophysical properties of salt marsh canopies — Quantifying plant stem flexibility and above ground biomass, *Coastal Engineering.* **100**, 48–57 (2015). doi: 10.1016/j.coastaleng.2015.03.009.

55. V. Vuik, H. Y. S. Heo, Z. Zhu, B. W. Borsje, and S. N. Jonkman, Stem breakage of salt marsh vegetation under wave forcing: A field and model study, *Estuarine, Coastal and Shelf Science.* **200**, 41–58 (2018). ISSN 0272-7714. doi: 10.1016/j.ecss.2017.09.028.

56. Z. Hu, P. W. J. M. Willemsen, B. W. Borsje, C. Wang, H. Wang, D. van der Wal, Z. Zhu, B. Oteman, V. Vuik, B. Evans, I. Möller, J.-P. Belliard, A. Van Braeckel, S. Temmerman, and T. J. Bouma, Synchronized high-resolution bed-level change and biophysical data from 10 marsh–mudflat sites in northwestern Europe, *Earth System Science Data.* **13**(2), 405–416 (Feb., 2021). ISSN 1866-3508. doi: 10.5194/essd-13-405-2021.

57. M. S. Fonseca, M. a. R. Koehl, and B. S. Kopp, Biomechanical factors contributing to self-organization in seagrass landscapes, *Journal of Experimental Marine Biology and Ecology.* **2**(340), 227–246 (2007). ISSN 0022-0981. doi: 10.1016/j.jembe.2006.09.015.

58. A. M. Folkard, Hydrodynamics of model Posidonia oceanica patches in shallow water, *Limnology and Oceanography.* **50**(5), 1592–1600 (2005). ISSN 1939-5590. doi: 10.4319/lo.2005.50.5.1592.

59. K. A. Moore, Influence of Seagrasses on Water Quality in Shallow Regions of the Lower Chesapeake Bay, *Journal of Coastal Research.* **2009**(10045), 162–178 (Nov., 2004). ISSN 0749-0208, 1551-5036. doi: 10.2112/SI45-162.1.

60. J. S. Weitzman, R. B. Zeller, F. I. M. Thomas, and J. R. Koseff, The attenuation of current- and wave-driven flow within submerged multispecific vegetative canopies, *Limnology and Oceanography.* **60**(6), 1855–1874 (2015). ISSN 1939-5590. doi: 10.1002/lno.10121.

61. R. A. Dalrymple, J. T. Kirby, and P. A. Hwang, Wave Diffraction Due to Areas of Energy Dissipation, *Journal of Waterway, Port, Coastal, and Ocean Engineering.* **110**(1), 67–79 (Feb., 1984). doi: 10.1061/(ASCE) 0733-950X(1984)110:1(67).

62. J. Dijkstra and R. Uittenbogaard, Modeling the Interaction between Flow and Highly Flexible Aquatic Vegetation, *Water Resour. Res.* **46**(12) (2010). doi: 10.1029/2010WR009246.

63. M. Baptist, V. Babovic, J. Rodríguez Uthurburu, M. Keijzer, R. Uittenbogaard, A. Mynett, and A. Verwey, On Inducing Equations for Vegetation Resistance, *Journal of Hydraulic Research.* **45**(4), 435–450 (July, 2007). ISSN 0022-1686. doi: 10.1080/00221686.2007.9521778.

Chapter 7

Incorporating Vegetation in Phase-Resolving Models – Macroscopic Approach

R. Divya[1] and V. Sriram[1,*]

[1]*Department of Ocean Engineering, Indian Institute of Technology Madras, Chennai, Tamil Nadu, India*
**vsriram@iitm.ac.in*

This chapter presents the mathematical and numerical approaches for wave interaction with porous structure. The porous structure can either represent a breakwater or a vegetation. Two modelling approaches such as microscopic and macroscopic are discussed. In macroscopic approach, the mathematical model is based on unified governing equation which is used for pure fluid and porous region. Additional drag force terms representing either the porous or vegetation is incorporated into the governing equation. Further, treating the interface between the pure fluid and porous region has been described. The implementation in phase resolving and phase averging models are reviewed and discussed.

Keywords: Wave-vegetation, wave-porous model, Morison's equation, phase averaging and phase resolving model, unified governing equation.

1. Introduction

Coastal hydrodynamics is a field of engineering which deals with the near shore processes in the ocean. Two main such process in the ocean is the waves and currents occurring due to various reasons. Sea waves or wind waves are mainly caused by wind blowing with an intensity for a long time and distance over the surface, while currents are caused by different reasons such as temperature gradient, bathymetric effect combined with

---

*Corresponding author.

waves etc., Sea waves are nothing but the harmonics caused over the sea surface, whereas currents are mass movement of water in a particular direction.

The sea waves which are formed deep in the ocean propagate towards the coast or shore. At initial stage, the wave heights are small which is not affected by the bottom topography, since the water depth is higher. As the waves propagate towards the coast, the water depth decreases leading to the bottom topography being felt by the wave. This in turn modifies the wave height and wave length to conserve the mass. Once the wave reaches the beach its height attenuates and wave breaks, dissipating all the energy. This is the basic phenomena happening in the ocean which was described in detail using a linear approximation by airy wave theory. The ocean wave are classified as per the water depth as deep water, intermediate water and shallow water depth, wherein the ration of water depth to wave length ranges from greater than 0.5, between 0.05 and 0.5 and less than 0.05, respectively. In the field of coastal engineering, when the waves enters from deep water to shallow region there are four different zones, namely offshore, nearshore, swash zone and back shore.

On the other hand, currents are just mass of moving water which will change the characteristics of the flows. Currents also interact with the wave and can change the wave height and wave length, thereby changing the intensity of the wave. Currents increase the phase of erosion in the shore when they move along the shore and are seasonal.

Coastal Structures are built on the coast to protect the living or non-living thing from the wrath of nature. These structures are built to slow down or prevent erosion, increase the access to the berthing of vessels in port and harbour and support coastal protection. These structures are more vulnerable since they are exposed to more of dynamic loads such as waves and currents than that of a land based structure. More than 70% of coastal structures are on shoreline to protect from probable damage due to events like storm surge. There are lot of coastal structures present based on the purpose they serve and are called as revetments, Bulkheads and Sea walls, Breakwater, Groins and Jetties, Coastal piers and Dikes. Revetments are structures parallel to the shoreline constructed mainly to absorb the energy of wave or currents incoming. Bulkheads are vertical structure which are constructed along the shoreline to hold the soil behind from sliding. Sea walls are coastal protection structure which ensures the land behind it is free from water entry due to storms or wave action. These are coastal defence structures mainly constructed to protect the land behind from erosion. Whereas breakwaters are the structures which are constructed perpendicular to the shoreline to maintain the wave tranquillity inside the port. These structures are constructed in such a way that the wave inside the port doesn't affect the moored vessels. Groins on the other hand are a structure which is constructed perpendicular to the shoreline to enhance the shoreline stability. These structures are effective when the longshore currents are higher. When these structures are placed in series the long-shore currents loses energy and deposits the sediment leading to increase in shore width. Jetties and piers are shore-normal structures commonly used for training navigation channels and stabilizing inlets. These are structures with protecting frame of pier and also used as a landing wharf. Finally, Dikes arc carthen structures which are constructed to keep the low land safe from elevated water levels. All these structure's main aim is to safeguard the coast from the wave, currents, storms etc., or to be specific reduces erosion.

2. Waves on permeable and impermeable coastal structures

The wave action over the coastal area plays a vital role in deciding the type of structure to be incorporated. If the main purpose is to reduce the wave overtopping then sea walls are constructed and if shore needs to be protected then breakwaters or vegetation can be constructed. All these structures are called as coastal defense structures since they protect the shore from wave action. All the theories have been arrived by assuming the structure to be rigid and impermeable which might not be the case in reality as most of the coastal structures are permeable in nature due to the usage of rubble mound stones or interlocking elements like tetrapod. In some situations, structure may be impermeable in nature like a rigid con-crete sea wall but even for these structures toe layers are present to avoid scouring, which are permeable in nature. Thus, wave forces obtained using the theory assuming the bed or the structure to be impermeable would always lead to a conservative result. If permeability is incorporated into the theory, it's difficult to solve the complex problem. Also the sea bed roughness and porosity is not taken into consideration while arriving at the wave deformation parameters such as shoaling, refractions, reflections. The sea bed interaction changes the wave kinematics locally due to the wave damping which will be significant for smooth bed when a wave travels for a longer distance. On the other hand, wave might cause significant bed deformation and stress leading to soil failure which in turn will change the forces acting on the pipelines buried on bed. Generally the wave is damped due to the presence of porous layer because of the drag resistance of the soil particles offered in the flow. In these cases, some energy is dissipated even before reaching the coastal structure.

Hence, with a permeable coastal structure there is always a higher rate of energy dissipation than that of impermeable one. Quantifying the amount of energy dissipation is required so that the effectiveness of each permeable structure is known and also the wave force actually hitting on the structure is arrived with which the structure can be economically designed. The simplest way to see the energy dissipation of a wave in a permeable structure is by checking the reflected and transmitted waves from structure. The ratio between the transmitted wave and incident wave gives the amount of energy dissipated. The energy dissipated can mainly depend on the type of permeable structure used. The structure used can either be hard measure or soft measure to reduce the wave energy. Hard measures are the one where breakwaters are used to reduce the energy while soft measure is the one where eco friendly solutions are used.

Wave vegetation interaction comes into picture when soft measures are suggested for coastal defense structure. Placing a vegetation belt over the coast not only increases the aesthetics of the coast but also it act as an environmental friendly solution. The aquatic lives can be enhanced over there and sustaining solution is obtained. When a vegetation belt is placed near a coast as a defense structure it experiences a lot of dynamic forces to which it has to withstand without failure. This leads to a field of wave vegetation interaction to understand whether the vegetation can sustain the dynamic forces acting on it through experimental and numerical approaches. Fig. 1 gives a picture of vegetation spread over the shore.

Fig. 1. Pine trees spread over the shore.

3. Modelling of porous structure

Wave porous structure interaction is commonly carried out using the experimental investigations. This should be performed before the structure is being installed at site, since the design formulae used to design the structure are approximate and may not holds good for all the scenarios. Thus, experimental investigations will give the visual observations on the performance of the designed structure configurations.

For varying the porosity of the structure or the dimension of the structure, experimentally lot of works has to be done. For instance, if a porous submerged breakwater is analysed for its effectiveness in wave damping or reflection characteristics, experimentally the layer arrangements of the breakwater has to varied for each parameters. Thus, experimental work becomes a time consuming and tedious process. Other disadvantage is that the facilities required for conducting the experiments are not easily available when it comes to replicating the site scenario. This leads to numerical studies on wave porous structure interaction, which can simulate the physical problem required in a short time and mostly economical than the experiments.

4. Mathematical model of the porous flow

Over the past decades, different porous flow models are suggested. These models can be classified based on the requirement of the problem at hand

  R. *Divya and V. Sriram*

i.e., macroscopic or microscopic approach. As the name implies, microscopic models are the one that consider model which have complete small scale flow features such as penetration of the fluid into the voids of the solid objects (such as stones) and the rigid body movement of such object. Whereas, macroscopic approach doesn't consider full feature modelling, however it's an approximate model that reflects the amount of dissipations. This has been pictorially represented in Fig. 2.

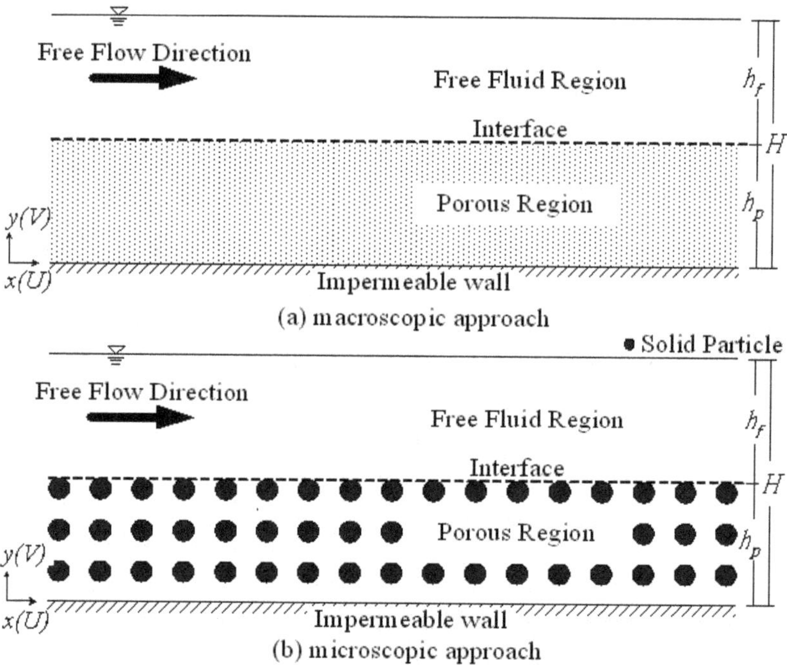

Fig. 2. Macroscopic and Microscopic porous model.

For the wave interaction with permeable coastal structures, later approach is computationally efficient than the former one, and it is good enough for the large scale flows. One of the simplest model for porous flow is proposed by Darcy (1856). Based on the flow in a porous region experiments, it was found that the velocity of flow is directly proportional to the pressure gradient. Further, this approach is valid only for laminar type of flows and suggested the following empirical equation,

$$-\nabla p = \frac{\mu}{K_p} u \tag{1}$$

Where ∇p is the pressure gradient inside a porous region, μ is the dynamic viscosity and K_p is a coefficient of permeability which indicates the porosity, connectivity etc., of the porous region and u is the discharge velocity inside the porous region. This model is the simplest of all and is accurate for steady flow with lower permeability in porous region. When the permeability is higher, the flow becomes turbulent and the velocity is over estimated. This leads to incorporation of non-linear terms by Dupit and Forcheimmer (1901) into the Darcy's suggestion.

$$-\nabla p = a\vec{u} + b\vec{u}|\vec{u}| \tag{2}$$

Where a is the linear drag coefficient as suggested by Darcy and b is the coefficient added to incorporate turbulent effects, which was defined by Ward (1964) as follows,

$$b = \frac{\rho C_f}{\sqrt{K_p}} \tag{2a}$$

Where ρ is the density of fluid and C_f is the non-dimensional coefficient. This model was found to be accurate in estimating the discharge velocity both in the case of laminar and turbulent flow. Later, Polubarinova and Kochina (1962) incorporated an inertia term to this model which was given as,

$$-\nabla p = a\vec{u} + b\vec{u^2} + c.\frac{\partial\vec{u}}{\partial t} \tag{3}$$

Where a, b and c are empirical coefficients and inertia term is the local acceleration term in Navier-Stokes equation for fluid flow. It should be noted that this inertia term indicates the inertia of fluid flow and not the inertia caused due to fluid and solid interaction or added mass effect. In order to incorporate the inertia force due to virtual mass effect due to fluid - solid interaction, Sollit and Cross (1972) extended the equation with two inertia components,

$$-\nabla p = \left(n_w \frac{\mu}{K_p} + n_w^2 \frac{\rho C_f}{\sqrt{K_p}} |\vec{u}|\right)\vec{u} + \rho\left(1 + \frac{1-n_w}{n_w} C_a\right)\frac{\partial\vec{u}}{\partial t} \tag{4}$$

                    R. Divya and V. Sriram

Where n_w is the porosity and C_a is the added mass coefficient. This governing equation of porous flows along with the fluid flow governing equation are solved and coupled at the interface in the macroscopic approach. The porous flow equations are incorporated into Navier-Stokes equation, which leads to equations such as VARANS (Volume Averaged Reynolds Average Navier Stokes) in which the velocity in the porous region is volume averaged. The flow in a porous medium is not the same throughout the medium as the pores and alignment of the solid particle varies in the spatial dimension. In order to approximate this, average value of flow properties are taken in the spatial dimension leading to volume averaged equation.

Hence, the velocity and pressure is replaced by two components namely the spatial averaged value and the spatially fluctuating quantity, leading to the following form of NS equations,

$$\frac{\partial \bar{u}}{\partial x_i} = 0 \tag{5}$$

$$\frac{1 + C_a}{n}\frac{\partial \bar{u}_i}{\partial t} + \frac{\bar{u}_j}{n^2}\frac{\partial \bar{u}_i}{\partial x_j} = -\frac{1}{\rho}\frac{\partial \bar{p}}{\partial x_i} + \frac{\vartheta}{n}\frac{\partial^2 \bar{u}_i}{\partial x_j \partial x_j} - \frac{1}{n^2}\frac{\partial \overline{u_i'' u_j''}}{\partial x_j} \tag{6}$$

Where the quantities with single dash represents the spatially averaged and with double quotes mean the spatial fluctuations. It can be noted that Equation (6) is arrived from the modification carried out in Darcy's law and incorporated directly into NS equation. Lin (1998) used this governing equation for the porous flow along with another set of governing equation for pure fluid flow, which was Reynolds averaged Navier Stokes equations (RANS) and used matching boundary at the interface.

4.1. *Drag and inertia components*

The dominant force acting in wave porous structure interaction is modelling the amount of drag resistance given by the solid particles. When the drag resistance in the flow is high it leads to higher energy dissipation. Over the past years, research works are concentrated in quantifying the amount of drag resistance provided by the solid particles to the fluid flow analytically, experimentally and numerically. Numerically drag resistance is taken into consideration by coefficients which were experimentally or analytically arrived. The coefficients can be broadly classified into three categories, namely the linear drag coefficient, non-linear drag coefficient and the transitional drag coefficient based on the type of flow. Thus, it is normal practice to decide based on Reynolds number. For low Reynolds number, the linear drag coefficient which represents the laminar flow becomes dominant and energy dissipation is only due to laminar flow over the solid particles. In the case of high Reynolds number, the non-linear drag coefficient which represents the turbulent flow becomes dominant and energy is dissipated mostly due to turbulence formed inside the solid particles.

The general forms of these coefficients were given below,

$$a = \alpha \frac{\vartheta_w (1 - n_w)^2}{n_w^3 D_{50}^2} \quad b = \beta \frac{1 - n_w}{n_w^3 D_{50}} \quad C_r = 1 + \frac{1 - n_w}{n_w} \gamma \text{ with } \gamma = 0.3 \quad (7)$$

Where, n_w is the porosity defined as the ratio between the volume of voids to the total volume (Le Mehaute, 1976), D_{50} is the mean soil particle size α and β are linear and non-linear constants need to be calculated from experiments.

Various constants for linear and non-linear drag have been reported in the literature from experiments. Carman (1937) used only linear constant and non-linear drag constant was assumed to be zero. Ergun (1952) based on experiments suggested a linear constant of 150 and non-linear constant of 1.75. Later, Engelund (1953) based on his experiments suggested linear constant 780-1500 and non-linear constant 1.8-3.6. Most widely adopted linear and nonlinear constant was suggested by Van Gent (1995) in which the linear constant of 1000 and non-linear constant of 1.1 was recommended. Liu (1999) suggested a linear and non-linear constant of 200 and 1.1. Ward (1964) has suggested an entirely different coefficient for drag forces based on coefficient of permeability and friction factor (C_f) of 0.51.

$$a = \frac{\vartheta}{K_p} \quad b = \frac{C_f n_w^2}{\sqrt{K_p}} \quad (8)$$

Karunarathna and Lin (2006) has included both linear and non-linear coefficient along with the transitional coefficient. The linear, nonlinear

 R. Divya and V. Sriram

and transitional coefficients are taken as 126, 0.51 and 4.5 respectively. The power of linear coefficient 'a' varies from one model to other, however the non-linear coefficient 'b' follows same functional form. Both 'a' and 'b' is lesser for Karunarathna and Lin (2006) than all other model, since they are accounting for an additional transitional term. These terms are basically derived from Morison equation, where soil particles are assumed to be collection of spheres. The inertia and drag force terms in porous layers are calculated based on this assumption. However, it should be noted that the drag force terms are arrived from steady flow assumptions. Once the drag force term is arrived, then based on Morison's equation over number of spherical soil particles, the coefficient of drag 'C_D' is taken based on the empirical formula suggested by Fair *et al.* (1968) which is as follows.

$$C_D = \frac{24.0}{R} + \frac{3.0}{\sqrt{R}} + 0.34 \tag{9a}$$

Karunarathna and Lin (2006) modified the above empirical equation to arrive at the linear, non-linear and transitional coefficients as given below.

$$C_D = c_1 \left(\frac{24.0}{R}\right) + c_2 \left(\frac{3.0}{\sqrt{R}} + 0.34\right)\left(1 + \frac{7.5}{KC}\right) \tag{9b}$$

Where, $KC = U_{max}T/n_w D_{50}$ is the Keulegan-Carpenter number in which U_{max} is the maximum water particle velocity, T is the wave period. The constants c_1 and c_2 were modification coefficients for laminar and turbulent flows and R is the Reynolds number. The last term was added to account for change of non-linear frictional force inside the porous medium due to oscillatory flows, thus slightly improving the drag force assumptions. Karunarathna and Lin (2006) suggested constant value of c_1 as 7.0 and c_2 as 2.0 assuming negligible mean turbulence effect inside the porous media. For arriving at the linear drag coefficient 'a', first term is taken into consideration for coefficient of drag. For arriving the non-linear drag coefficient 'b' term $0.32\,c_2$ is taken as coefficient of drag and for the transitional coefficient 'c' the square root term $c_2\left(\frac{3.0}{\sqrt{R}}\right)$ is taken as coefficient of drag. Since the turbulence effect inside the porous medium is negligible, the KC term is neglected in deriving the coefficients.

The idea of the porous flow model can be extended for wave vegetation interaction considering the fact that the vegetation acts as an porous medium and damps the incoming wave energy. The following section explains the wave vegetation model and how similar it is to wave porous model for solving it numerically.

5. Modelling of vegetation

Generally large scale domains are modeled numerically to arrive at the wave vegetation interaction. Similar to porous structure modelling, numerical model can either be macroscopic or microscopic. Macroscopic model is a one in which the vegetation belt is considered as an damping zone as a whole while microscopic model is where each and every stem is modeled and interaction is studied. The limitation of the macroscopic model is that the wave force acting on the individual vegetation stem cannot be arrived whereas the limitation of microscopic model is that the computation cost is high. When the problem under consideration is only the wave characteristics then macroscopic model is preferred. There are two approaches in modeling wave vegetation (a) an additional resistance force term is added to the momentum equation as derived by Su and Lin (2005) (b) conservation of energy is taken into consideration for modeling and an energy dissipation damping term is added to the final wave attenuation equation as done by Darlymple *et al.* (1984). When the momentum equation is considered the flow properties are better obtained than considering energy conservation equation. Generally, a momentum equation in a macroscopic level is modeled to observe the wave vegetation interaction.

RANS equation is used to model the wave vegetation interaction by adding forces such as drag and inertia due to vegetation in the original NS equations. Su and Lin (2005) considered the vegetation to be a collection of uniform, rigid and smooth cylinder in a control volume (V), the inertia and drag force over the cylinder is obtained by Morison's equation as,

$$F_D = \frac{1}{\rho V_f}\left(\frac{1}{2}\rho C_D A \frac{\bar{u}_c}{n_v}\frac{\bar{u}_i}{n_v}\right) \tag{10}$$

$$F_I = \frac{1}{\rho V_f}\left(\rho C_m V_p \frac{1}{n_v}\frac{\partial \bar{u}_i}{\partial t}\right) \tag{11}$$

Where F_D is the drag force and F_I is the inertia force obtained for a single-cylinder with a cross-sectional area A perpendicular to the flow, ρ the water density, V_f the volume of fluid in the control volume, V_p the volume of a single-cylinder, n_v the porous rate, $\bar{u}_c$ the characteristic velocity $= \sqrt{\overline{u_i u_j}}$ in which i and j are the direction of flow, C_D drag force coefficient and C_m inertia force coefficient. Equation (10) and (11) is then applied for N number of cylinders present in the control volume, and we get

$$F_D = \frac{1}{\rho V_f}\left(N\frac{1}{2}\rho C_D A \frac{\bar{u}_c}{n_v}\frac{\bar{u}_i}{n_v}\right) \tag{12}$$

$$F_I = \frac{1}{\rho V_f}\left(N\rho C_m V_p \frac{1}{n_v}\frac{\partial \bar{u}_i}{\partial t}\right) \tag{13}$$

Now, Equation (12) is taken to arrive at the drag force for a group of cylinders representing the vegetation. Multiply and divide V_p to Equation (12) and apply the relation $NV_p = V_s$ and $n_v = V_f/(V_f + V_s)$, where V_s is the total volume of solids in the control volume V. Then, incorporate the volume and area of a cylinder of diameter 'D_v' and height 'h' in the arrived equation to get the final form as below,

$$F_{D\_c} = \frac{2}{\pi}\left(\frac{(1-n_v)}{n_v{}^3}C_D \frac{1}{D_v}\bar{u}_c\bar{u}_i\right) = f_{d\_c}\bar{u}_c\bar{u}_i \tag{14}$$

In case, if the vegetation is assumed to be a collection of strips of width 'b_v' and thickness 't_v' then the value V_s and A is changed accordingly in the Equation (12), and the final form is as follows,

$$F_{D\_s} = \frac{(1-n_v)}{n_v{}^3}\left(\frac{C_D}{2.t_v}\bar{u}_c\bar{u}_i\right) = f_{d\_s}\bar{u}_c\bar{u}_i \tag{15}$$

To get a generalized form for the porous rate 'n', a control volume of dimension $\Delta x.\Delta y.\Delta z$ is taken, where N number of the cylinder is present. Applying the relation of the porous rate $n_v = V_f/(V_f + V_s)$ in the generalized form for cylinder (Eqn. (16)) and strips/square (Eqn. (17)) vegetation, the porous rate can be given as,

$$n_v = 1 - \left(\frac{\pi D_v^2}{4} N_D\right) \tag{16}$$

$$n_v = 1 - (b_v t_v N_D) \tag{17}$$

Where N_D, the vegetation density, which is the number of vegetation stems per unit area.

Inertia force for a group of cylinder/strips is arrived from Eqn. (13) by applying the relation $V_p = V_s$, the final form is as below,

$$F_I = C_m \frac{(1 - n_v)}{n_v^2} \frac{\partial \overline{u}_i}{\partial t} \tag{18}$$

The arrived inertia and drag force term is added to the original momentum equation to get the interaction of vegetation.

5.1. *Phase resolved and Phase averaged wave vegetation model*

In a broader view, wave vegetation interaction can be modeled either using a phase resolved wave model or phase averaged wave model. In simple terms, a phase averaged wave model is a one which uses the wave spectrum as an input and not the individual wave. It is usually developed using wave action conservation equation or mild slope equation in which the wave attenuation due to vegetation is considered. Generally, wave vegetation interaction is simulated using phase averaged model such as SWAN. Chen *et al.* (2011) used SWAN model and carried out the wave transformation over plants. Suzuki *et al.* (2012) studied the wave dissipation in a vegetation field in a full spectrum based on SWAN model. Maa *et al.* (2002) used a RIDE model which consist of mild slope equation and studied the wave propagation over the vegetation zone by adding a vegetation module in the original equation. Cai and Zhu (2014) studied the vegetation induced wave damping by weakly non-linear RIDE model. Tang *et al.* (2015) used a parabolic mild slope equation and studied wave transformation in rigid vegetation by adding a vegetation induced wave dissipation module.

 R. Divya and V. Sriram

The advantage of the phase averaged model is that there is no requirement of minimum number of computation points per wavelength which is required in phase resolved models. The phase averaged model takes up lot of assumptions than the phase resolved model. The phase resolved models are Reynolds averaged Navier stokes (RANS) equation, Boussinesq and shallow water equations.

5.2. *Boussinesq equation*

The phase resolving models used widely for wave vegetation interaction are based on Boussinesq equations. These are two-dimensional horizontal (2DH) models which can capture 3D physics using a 2D mesh. This is achieved either by integrating the governing equations over the water depth, or by evaluating the equations at a tuned depth, with various levels of assumptions. Over the years, a number of Boussinesq-type equations have been developed with varying regions of applicability and complexity based on the assumptions, non-linear and dispersive terms. A details review for these equations is presented in Brocchini *et al.* (2013) and Yang *et al.* (2020). These equations can be developed into power tools for simulating wave propagation over large domains with uneven topography.

For improved practical application in the coastal environment, the Boussinesq equation models often incorporate flow through porous structures such as breakwaters or vegetation, based on Darcy velocity theory, with additional laminar ($\propto u$) and turbulent ($\propto |u|u$) drag resistance terms. A brief review on the constant of proportionality for these terms applied to the Boussinesq equations is presented in Burchart *et al.* (1995). Early simulations for diffraction around porous breakwaters were limited to thin structures, Yu *et al.* (1997). Improvements in stability of the Boussinesq models lead to further applications in interaction of solitary waves with porous detached breakwater in Lynett *et al.* (2000). The porous breakwater approach was further modified for modelling interaction of waves with vegetation. Augustin *et al.* (2009) analyzed wave transformation in the presence of vegetated water using modified Boussinesq equation by increasing the friction factor due to vegetation. Kuiry *et al.* (2011) used an one dimensional Boussinesq equation to simulate wave attenuation due to vegetation. Similarly lot of wave vegetation models were developed using Boussinesq equation such as Huang *et al.* (2011), Wang *et al.* (2015), Zhang *et al.* (2012), Yao *et al.* (2015) etc., All the above mentioned model analyzed the wave damping and the influence of vegetation dimensions were not much studied. Yang *et al.* (2018) numerically studied the effect of wave attenuation due to vegetation configurations and also its response for various wave period, water depth and wave height.

We present brief description of the implementation of flow through porous structures and flow through vegetation in two Boussinesq models.

5.3. *Flow through porous breakwater in FEBOUSS*

FEBOUSS is a general-purpose finite element model for weakly non-linear weakly dispersive depth integrated form of Boussinesq equation developed at IIT Madras (Agarwal *et al.*, 2022). The governing equations are of the following form,

$$\frac{\partial \eta}{\partial t} + \frac{\partial}{\partial x}\frac{P}{\lambda} + \frac{\partial}{\partial y}\frac{Q}{\lambda} = 0 \tag{19}$$

$$\frac{\partial P}{\partial t} + \frac{\partial}{\partial x}\frac{P^2}{\lambda d} + \frac{\partial}{\partial y}\frac{PQ}{\lambda d} + \lambda g d\frac{\partial \eta}{\partial x} + \lambda f_l P + \lambda f_t \frac{P}{d}\sqrt{P^2 + Q^2} + \Psi_x = 0 \tag{20}$$

$$\frac{\partial Q}{\partial t} + \frac{\partial}{\partial x}\frac{PQ}{\lambda d} + \frac{\partial}{\partial y}\frac{Q^2}{\lambda d} + \lambda g d\frac{\partial \eta}{\partial y} + \lambda f_l Q + \lambda f_t \frac{Q}{d}\sqrt{P^2 + Q^2} + \Psi_y = 0 \tag{21}$$

Here 'η' is the surface elevation, 'P' and 'Q' are the depth integrated velocities along the X and Y axes, 'h' is the still-water depth, '$d = h + \eta$' is the total water depth, 'g' is the acceleration due to gravity and Ψx and Ψy are the high order dispersive terms. Here λ is the local porosity, where $\lambda = 1$ corresponds to a fluid region without porosity. The laminar and turbulence drag terms are given by $\lambda f_l P$ and $\lambda f_t \frac{P}{d}\sqrt{P^2 + Q^2}$, where f_l and f_t are the respective coefficients given by the following empirical relations,

$$f_l = \alpha_0 \frac{(1-\lambda)^3}{\lambda^2}\frac{\nu}{s^2} \;,\; f_t = \beta_0 \frac{(1-\lambda)}{\lambda^3}\frac{1}{s} \tag{22}$$

                    R. Divya and V. Sriram

Here, 'v' is the kinematic viscosity of water, 's' is the characteristic size of the stone and 'α_0' and 'β_0' are empirical constants with recommended range of 780–1500 and 1.8–3.6, respectively, based on Engelund (1953). The FEM model is built on unstructured triangular mesh with mixed linear and quadratic shape functions, with high order RK4 time-stepping. The details regarding the governing equations and the numerical formulation can be found in Agarwal *et al.* (2022), with specific modifications for flow through porous structures in Agarwal *et al.* (2019).

The model was used for studying the partial reflection and partial transmission of a solitary wave through porous breakwater for a range of wave-heights, stone size and breakwater width in Agarwal *et al.* (2019). Fig. (3) shows the phenomenon of splitting of a solitary wave into a reflected and a transmitted wave when interacting with a porous break-water. The heights of the reflected and transmitted waves depend upon the characteristics of the breakwater, as shown in Fig. (4). As most Boussinesq models are built on potential flow assumptions, additional terms should be included for wave dissipation attributed to turbulence, even outside the porous region.

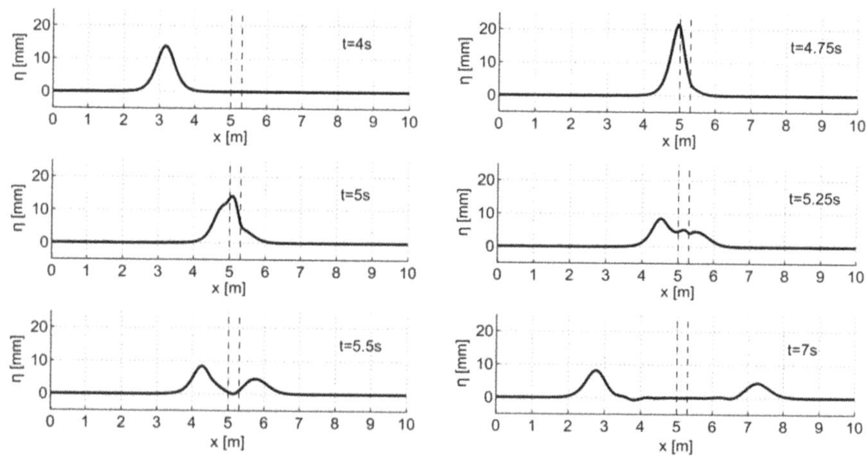

Fig. 3. Snapshots of surface elevation at different time instants depicting partial transmission and partial reflection of solitary wave through the porous breakwater (source Agarwal *et al.*, 2019).

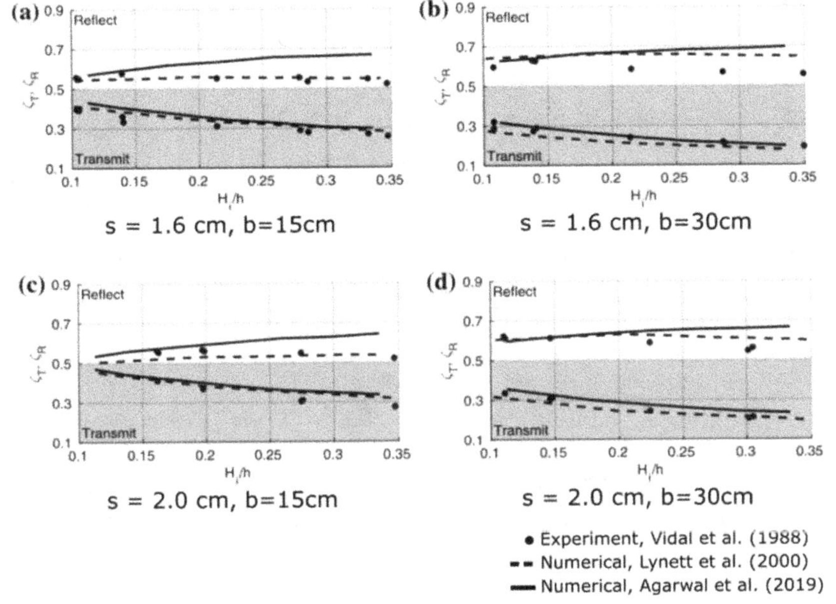

Fig. 4. Graphs showing estimated transmission and reflection coefficient for 1D solitary waves through porous breakwater; shaded region shows transmission

5.4. *Wave interaction with vegetation in COULWAVE*

COULWAVE is a finite difference model for a highly nonlinear weak dispersive form of the Boussinesq equations. The governing equation are of the following form,

$$\frac{\partial \zeta}{\partial t} + \nabla . [(\zeta + h)u_\alpha] + B_c = 0 \tag{23}$$

$$\frac{\partial u_\alpha}{\partial t} + g\nabla\zeta + \frac{1}{2}\nabla|u_\alpha^2| + B_m + R_f - R_b + R_{veg} = 0 \tag{24}$$

Where ζ is the free surface elevation, h is the water depth, u_α is the reference horizontal velocity vector at tuned depth of $Z_\alpha=0.531h$. B_c and B_m are the higher order non-linear terms g the acceleration due to gravity. Complete details regarding the governing equations and the numerical formulation for COULWAVE can be found in Lynett *et al.* (2002). The model was extended for modelling flow wave interaction with vegetation in Yan *et al.* (2018). R_f is the term accounting for bottom friction, R_b accounts for wave breaking and R_{veg} is the term for vegetation effects. Since the diameter of vegetation small compared to wave height, inertia force is neglected and R_{veg} is defined as,

$$R_{veg} = \frac{1}{2}\rho C_D N b_v u_c |u_c| \tag{25}$$

Where ρ is the density of fluid, C_D is the coefficient of drag, N is the vegetation density b_v is the vegetation width and u_c is the velocity averaged over the height of vegetation layer which is given by,

$$u_c = u_\alpha \left(\frac{h_v}{d}\right)^{1/2} \eta_v \tag{26}$$

Where h_v is the submerged height of vegetation and η_v is the coefficient based on analytical equations. This study found that the vegetation configuration also affects the wave energy attenuation.

Thus, Boussinesq equation can be used to incorporate non linearity in the wave vegetation interaction and also combined effects if wave breaking and friction can be taken care.

6. Mesh based method in wave vegetation interaction (RANS)

There are various mesh based methods like Finite element, Finite Volume, and Finite difference method used to numerically solve the RANS equation. Li *et al.* (2010) used a σ coordinate 3D model to investigate a wave-induced mixing process in a vegetation field. Zhang *et al.* (2013) used the finite difference method to solve the Reynolds Averaged Navier Stokes (RANS) equation with vegetation forces. Ma *et al.* (2013) used a non-hydrostatic RANS model based on NHWAVE to study the turbulent mixing, surface wave attenuation, and near-shore circulation induced by vegetation. Marsooli *et al.* (2014) investigated the wave attenuation with the 3D RANS model and solved by the finite volume method. Maza *et al.* (2016) used OpenFOAM to investigate the solitary wave attenuation due to vegetation patches. Chen *et al.* (2016) used a commercial CFD code FLUENT to study wave-vegetation interactions and its forces. All the above literature was based on mesh methods in the Eulerian framework, wherein meshing complex geometry becomes difficult. The Eulerian framework also requires the finer mesh to capture extreme event interaction with vegetation, which increases the computational cost.

7. Meshless methods (VARANS)

In the Eulerian formulation using mesh based schemes, special treatment is required for convective terms, fine mesh near the point of interests, free surface tracking for large free surface deformation in the case of two-phase flow. In Lagrangian formulations, the mesh or particle will move with the fluid particles. The large deformations may be modeled effectively than in the Eulerian approach.

Meshless methods are quite popular because of mesh being neglected and particles being used. The main advantage of this type of numerical method is that it has an easier adaptive analysis than the conventional mesh methods as well as it avoid mesh diffusion. As far as wave vegetation interaction is concerned lot of meshless methods are developed. The vegetation can be considered as a porous structure since the model governing equation is macroscopic and individual vegetation is not modeled generally.

Some of the notable works in wave porous structure interaction were being done by Shao (2010), Akbari *et al.* (2013) and Ren *et al.* (2014, 2016). Shao (2010) solved two separate equations for fluid and porous region and coupled the flow property at the interface. The governing equation used was based on Ward's (1964) suggestion with friction factor and coefficient of permeability. The inertia coefficient is taken as unity. A two-step semi implicit method based on Incompressible Smooth Particle Hydrodynamics (ISPH) was used to solve the governing equation. The model was validated for solitary and regular waves after which it was applied in breaking wave overtopping over a porous breakwater. In this case no turbulence was taken into account and macroscopic flow properties were only analysed.

On the other hand, unified governing equation was used by Akbari *et al.* (2013) and solved using Incompressible Smooth Particle Hydrodynamics in Porous media (ISPHP). This unified governing equation adopted was based on Dupit and Forcheimmer's (1901) suggestion of linear and non-linear drag coefficient and Brinkmann's (1947) suggestion for effective viscosity coupled to the Navier Stokes equation along with inertia coefficient. The constants for linear, non-linear and inertia were taken from Van Gent's suggestion. This unified governing equation was suggested by Solit and Cross (1972) into the momentum equation. Akbari *et al.* (2013) used a two-step projection method in solving the governing equation where the pressure term is neglected and intermediate velocity is arrived after which the pressure poisson equation is used to get the pressure value implicitly. In the porous region, pressure poisson equation has all the term related to porosity in the right hand side which makes the stiffness matrix to be modified. This pressure poisson equation is thus solved based on ISPH. In this model, accuracy was obtained with 7-8 particles along the vertical between the wave crest and trough.

Recently, Ren *et al.* (2016) used a Weakly Compressible Smooth Particle Hydrodynamics (WCSPH) in which the fluid particles where assumed to be weakly compressible. The given field function and their gradient were arrived based on the kernel function and particle approximation. This kernel function is arrived based on a smoothing length for each and every particle with which a weight function is arrived. Different particle spacing is taken for pure fluid and porous region. The smoothing length for pure fluid is taken as 1.3 dx and for porous region it is 1.3 $dx/\sqrt{n_w}$. An explicit symplectic scheme for time marching was used to update the field variables, where the particle position is updated at the middle of the time step using the previous time intrinsic fluid velocity and also at the end of the time step using the updated intrinsic velocity. This method shows fair agreement with experimental results but the computation time will be higher due to difference in particle spacing and smoothing length.

8. Interface treatment between vegetation and wave

The interface between the fluid domain and the vegetation domain can either be strongly coupled or weakly coupled. Coupling in the interface means the physical variable is transferred from the fluid to vegetation domain. In a lagrangian framework, the particle keeps moving with time so the interface should be perfectly captured at every time step for maintaining continuity. When a vegetation is modeled in an macroscopic scale, it can be treated as a porous medium where the wave gets attenuated.

A Lagrangian framework leads to particle movement from fluid to porous region or vice versa during the course of the simulations; this further increases the importance of the interface treatment. Akbari *et al.* (2013) for this reason used background nodes with porosity information incorporated in it. This background mesh is stationary and has porosity value as unity in pure fluid region and given porosity value in porous region. A finer background mesh with porosity information is defined with which at each and every time step the fluid particle gets the information regarding the porosity from the background nodes using SPH inter-polation. For this interpolation, a smoothing length of mean porous grain diameter is taken into account.

Using this idea of background mesh, Ren *et al.* (2014) suggested a transition zone of length equal to smoothing length. This imaginary transition zone is placed between the pure fluid and porous region and it is divided into two equal halves the fluid part and porous part. At each computational time step, the velocity of the particles near to the transition zone in the fluid part is calculated using the fluid particles only, after truncating the porous particles. Similarly, the velocity of particles near to the transition zone in porous part is calculated based on only porous particles. Whereas, for the particles that lies in in the transition zone, the velocity is estimated by averaging the velocity of a nearby fluid and porous region particle. By doing this the velocity is smoothly transferred in the interface region, using this updated velocity, the pressure and densities are updated maintaining the continuity.

Further in works of Ren *et al.* (2016) with a unified governing equation, the concept of background nodes and transition zone was put together and a transition zone of linearly varying porosity value was suggested. The background nodes were initiated with a spacing of $dx/3$, where dx is the initial particle spacing. Here the transition zone was taken with a width of two times the smoothing length, where the porosity values of the back-ground nodes linearly varies from unity to the given porosity value. For interpolating from background nodes, SPH interpolation was used at every time step. Here the smoothing length was taken as the half of the mean porous grain diameter.

Divya and Sriram (2017) developed a wave vegetation model based on IMLPG_R in which background nodes (BG) has been considered. The porous rate n_v is provided based on the vegetation density and dimensions. The n_v value is interpolated using Simplified Finite Different Interpolation (SFDI in Ma, 2008) from the BG nodes for the interface region. The n_v value varies linearly in the transition zone between fluid and solid, and this will lead to smooth pressure estimates. The transition zone width is taken as 2 times the initial particle spacing based on the analysis previously carried out in Divya and Sriram (2017).

9. Mathematical model for wave porous and wave vegetation interaction using IMLPG_R

A Cartesian coordinate system is used with XY plane with X axis being rightward positive and Y axis being upward positive. The computational domain of length 'L$_t$', water depth 'd' and porous region depth as (d$_p$) is shown in Fig. 5 The fluid is assumed to be incompressible and governed by the following unified governing continuity and momentum equation for both fluid and porous region.

$$\nabla \cdot \overrightarrow{u_D} = 0 \tag{27}$$

$$\frac{C_r}{n_w} \cdot \frac{D\overrightarrow{u_D}}{Dt} = \frac{-1}{\rho_w}\vec{\nabla} P + \vartheta_{eff}\nabla^2 \vec{u}_D - \beta(\beta_p) - (1-\beta)(\beta_v) + \vec{g} \tag{28}$$

Where, D/Dt is the substantial derivative, C$_r$ is the inertia coefficient; n_w is the porosity; $\overrightarrow{u_D}$ is the Darcian velocity; P is the pressure; υ_{eff} is the effective viscosity based on Brinkman; g is acceleration due to gravity; ρ_w is the apparent density. β value is taken as 1 if porous structure is used and 0 if the vegetation is used. Equation (31a) gives the drag force for porous structure (β_p) and (31b) gives the drag force considered for vegetation structure (β_v).

The Darcian velocity $\overrightarrow{u_D}$ is basically arrived from the assumption of volume averaged velocity inside the porous region. For this, the intrinsic fluid velocity is integrated over the volume of fluids and divided by total volume of porous region. Since the porosity is defined as the ratio between volume of voids and total volume, the final expression lead to the following,

$$\overrightarrow{u_D} = n_w \vec{u}_f \tag{29}$$

$$\vartheta_{eff} = \frac{\vartheta}{n_w} \tag{30}$$

$$\beta_p = a\vec{u}_D + b\vec{u}_D|\vec{u}_D| + c\vec{u}_D\sqrt{\vec{u}_D} \tag{31a}$$

$$\beta_v = f_d\sqrt{|\vec{u}_D.\vec{u}_D|}.\vec{u}_D \tag{31b}$$

Where a, b and c is the linear, non-linear drag coefficients and transitional coefficient, respectively.

The drag force f_d for the vegetation is arrived based on the Equation (14) and (15) for cylinder and strips, respectively and given as,

$$f_{d\_c} = \frac{2}{\pi}.C_D.\frac{(1-n_v)}{n_v^3}.\frac{1}{D_v} \tag{32}$$

$$f_{d\_s} = C_D.\frac{(1-n_v)}{n_v^3}.\frac{1}{2t_v} \tag{33}$$

The coefficient C_D is tricky since the value has to be for a group of a cylinder and not for a single cylinder. Further, this value is not the same for the cylinder and strip type. Normally, it is assumed that C_D a function of water depth, wave period, wave height, type of vegetation, and is taken as a constant for the whole region of vegetation.

The Lagrangian form of kinematic and dynamic boundary conditions on the free surface are as follows

$$\frac{d\vec{r}}{dt} = \overrightarrow{u_f} \tag{34}$$

$$P = 0 \tag{35}$$

Where $\vec{r} = x\vec{i} + y\vec{j}$ is the position vector with respect to origin of coordinate system and $\overrightarrow{u_f}$ is the fluid velocity. The atmospheric pressure is taken as zero at the mean free surface.

On the solid wall boundaries such as bottom wall and structure the following boundary condition is applied which is arrived from the momentum equation,

$$\vec{n}.\overrightarrow{u_D} = \vec{n}.\overrightarrow{U_D} \tag{36}$$

$$\vec{n}.\nabla P = \rho_w.\vec{n}(\vartheta_{eff}\nabla^2\vec{U}_D - \beta(\beta_p) - (1-\beta)(\beta_v) + \vec{g} - \overrightarrow{\dot{U}_D}.\frac{C_r}{n_w}) \tag{37}$$

Where, $\overrightarrow{\dot{U}_D}$ = solid boundary acceleration; $\overrightarrow{U_D}$ = solid boundary velocity; $\overrightarrow{u_D}$ = fluid velocity. This Eqn. (37) is arrived from momentum equation Eqn. (28) in which the pressure term is taken to right hand side and substantial derivative to the left hand side assuming it to be the solid boundary acceleration.

The constant C_r varies for porous and vegetation. C_r is taken as 1 for porous structure meaning the inertial forces due to porous structure are less and for vegetation C_r is due to the added mass effect in the vegetation and local acceleration of the fluid. This is given as follows,

$$C_r = 1 + C_m\frac{(1-n_v)}{n_v} \tag{38}$$

Where C_m, the added mass coefficient.

10. Numerical modeling: wave-porous and vegetation structure

The governing equation along with the boundary condition will be solved based on Chorin (1967) time split algorithm, which is given below,

(a) In the first step, the intermediate velocity and position is obtained by neglecting the pressure term from governing Equation (28) leading to,

$$\vec{u}_D^* = \vec{u}_D^n + \frac{n_v * \Delta t}{C_r}\left[\vartheta_{eff}\nabla^2\vec{u}_D - \beta(\beta_p) - (1-\beta)(\beta_v)\right] \tag{39}$$

$$r^* = \vec{r}^n + \vec{u}_D^*\Delta t \tag{40}$$

(b) Then the Pressure Poisson Equation (PPE) is arrived and solved using the intermediate velocity based on the semi-implicit method,

$$\nabla^2 P^{n+1} = \frac{C_r * \rho}{\Delta t * n_v} . \nabla \vec{u}_D^*$$ (41)

(c) The pressure gradient at $n+1$ timestep is used to get the Darcian velocity as follows,

$$\vec{u}_D^{**} = \frac{-\Delta t * n_v}{\rho * C_r} \nabla P^{n+1}$$ (42)

$$\vec{u}_D^{n+1} = \vec{u}_D^* - \frac{\Delta t * n_v}{\rho * C_r} \nabla P^{n+1} + \vec{g}.\Delta t$$ (43)

(d) By using the Darcian velocity, the fluid velocity is obtained based on the relation given in Equation (42) leading to the following equations to obtain the new position and velocity,

$$\vec{u}_f^{n+1} = \frac{\vec{u}_D^{n+1}}{n_v}$$ (44)

$$\vec{r}^{n+1} = \vec{r}^n + \vec{u}_f^{n+1}\Delta t$$ (45)

(e) The simulation is repeated until the desired time.

The stability of the method is enhanced by updating the coordinate based on the minimum pressure gradient, see details in Sriram and Ma, (2012).

10.1. *Weak formulation of PPE*

Eqn. (37) can be solved either by using the strong form or weak form formulations based on mesh or mesh-free methods. Our research group preliminary used a mesh-free weak form method namely Improved Meshless Local Petrov-Galerkin with Rankine source (IMLPG_R) to solve the PPE. As the name suggests, this method doesn't require any mesh to discretize the computation domain, and particles are used instead, hence the term 'Meshless'. The interpolation is done for the local domain and not for the global domain, hence the term 'Local.' In the weak form, different test and trial functions are used, so the term 'Petrov-Galerkin' and the test function here we used is based on the Rankine source solution, as given below,

$$\varphi = \frac{1}{2\pi} * ln\left(\frac{r}{R_I}\right) \tag{46}$$

Where R_I , the radius of influence of the support domain. φ becomes zero at the boundary and satisfies Laplace condition $\nabla^2\varphi = 0$ in domain Ω_I except for the center and boundary $\partial\Omega$. This test function is then multiplied on both sides of PPE. The PPE is solved similar to the previous work done (Divya and Sriram, 2017) with the only difference is that instead of porosity n_w porous rate n_v is used in the current work which leads to the final general form,

$$[K_{IJ}].\,[P] = [F_I] \tag{47}$$

$$K_{IJ} = \begin{cases} \dfrac{1}{2\pi R}\displaystyle\int \phi_j(x).\,ds - \phi_j(x)\ for\ Inner\ fluid\ particle \\[2ex] \vec{n}.\,\nabla\phi_j(\vec{x})\ for\ Solid\ Boundary \\[2ex] \phi_j(x)\ for\ free\ suface\ particles \end{cases} \tag{48}$$

$$F_I = \begin{cases} \dfrac{c_r * \rho}{n_v(i) * \Delta t} * \dfrac{1}{2\pi}\displaystyle\int \vec{u}_D^*.\,d\Omega\ for\ Inner\ fluid\ Particle \\[2ex] \dfrac{c_r * \rho}{n_v(i) * \Delta t}\vec{n}.\,\vec{u}_D^*\ for\ solid\ boundary\ particle \\[2ex] P_{atm}\ for\ free\ surface\ particle \end{cases} \tag{49}$$

The advantage of this local weak form is that no pressure gradient and intermediate velocity gradient term is used in finding the pressure value, which enhances the accuracy of the pressure value thus obtained. Further, the influences of the porosity, inertia coefficients affect only on the right-hand side matrix, and the left-hand side matrix depends upon the coordinates of the particles, and thus will not affect the original matrix solvers implemented in MLPG based on Gauss-Seidel and GMRES.

11. Validation of strip-type submerged vegetation

The model is validated for strip type of vegetation in submerged conditions, which was experimentally investigated by Asano *et al.* (1992). The experiments were conducted in a wave tank of length 27m, width 0.5m, and depth 0.7m. The vegetation was represented by flexible polypropylene strips having a specific gravity of 0.9. The strips have a height of 0.25m, a width of 0.052m, and a thickness of 0.03mm in a water depth of 0.45m. These strips were placed for a length of 8m with a different density for each test case. Four wave-gauges were placed along the length of the vegetation, and the wave height attenuation of the regular wave was recorded.

Fig. 5 shows the computation domain having a length of 27m in a water depth of 0.45m. The particle spacing was taken as 20 particles across the water depth, and the time step was chosen as 0.01s. Table 1 gives the simulated test conditions in the present numerical model. The C_d and N_d are provided in Asano *et al.* (1992).

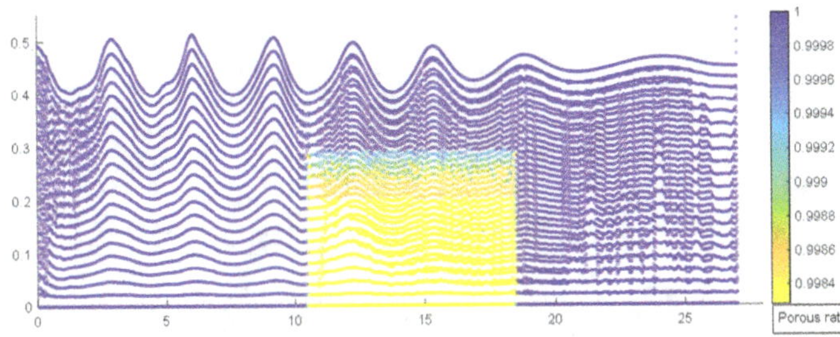

Fig. 5. Computation domain for strip vegetation with a porous value under regular waves (Divya and Sriram, 2020).

Table 1. Experimental cases taken from Asano (1992) for numerical simulation.

Case no	Incident wave height H_i (*m*)	Time Period T (*s*)	Coefficient of drag C_d	Vegetation density N_d (*stems/m²*)
1	0.0360	2.00	0.18	1490
2	0.1001	1.67	0.13	1100

Further, the C_m value set to 2 for all the test cases. The wave height at four places starting from the vegetation patch with 2m intervals is taken and compared for all the four cases for the experimental, numerical, and analytical results. The numerical results are from Marsooli and Wu (2014), in which the 3D equation based on RANS is solved using the Finite volume method. The results are shown in Fig. 6, and good agreement is found by the proposed numerical model based on macroscopic approach. The accuracy level of the proposed 2D numerical model is almost similar to that of Marsooli and Wu (2014).

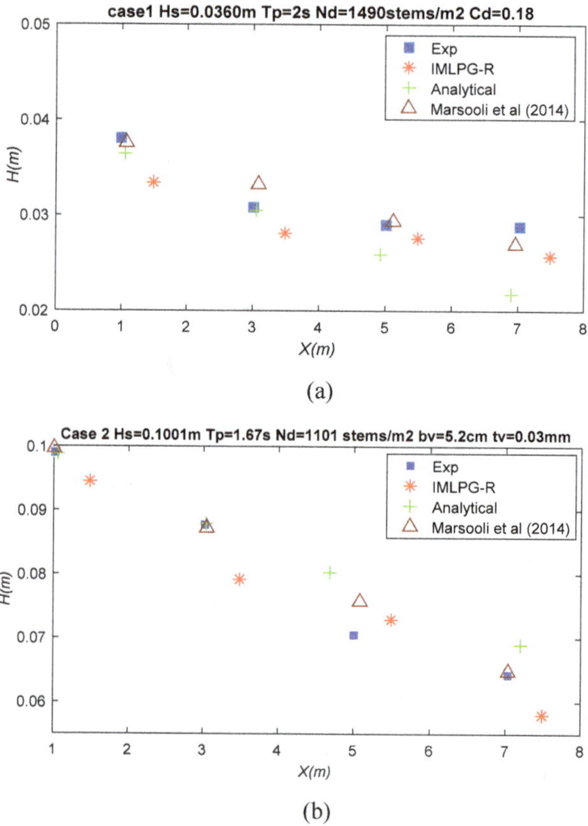

Fig. 6. Comparison of experimental (Asano, 1993) results, Analytical, Numerical results from Marsooli and Wu(2014) and IMLPG_R results for (a) case1 (b) case2 (Divya and Sriram, 2020).

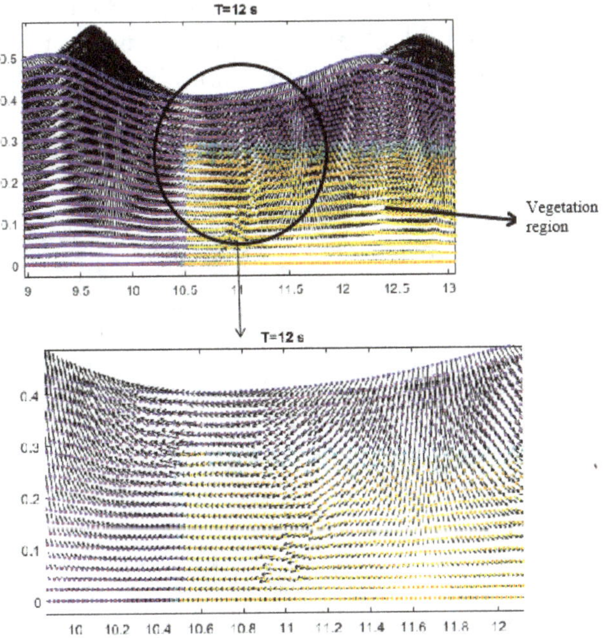

Fig. 7. Velocity variations through the depth in pure fluid and vegetation region (Divya and Sriram, 2020).

The pressure and the velocity variation in the wave over a pure fluid region and submerged vegetation region are given in Fig. 7. It is evident that at the instant, when the wave enters the vegetation region, the velocity vector reduces in magnitude and influence across the vegetation region is less.

12. Numerical simulation of cylindrical vegetation as emerged and submerged

The performance of cylindrical vegetation model type using the experimental results from Augustin *et al.* (2009) is analysed. Monochromatic waves were analyzed in a wave flume of length 30.5m, width 0.9m, and depth 1.2m. The vegetation was placed approximately 13.1m from the wavemaker and was kept for 6m length. Rigid vegetation is modeled by using cylindrical wooden bowels of diameter 0.012m and height 0.3m with a vegetation density of the order 162 stems/m^2. The following cases were simulated as given in Table 2, and the wave attenuation is calculated in four places at 3m, 5m, and 6m from the start of vegetation.

Table 2. Test cases simulated from Augustin *et al.* (2009).

Type of vegetation	Wave height H_i (m)	Time period T (s)	Coefficient of Drag C_d	Inertia Coefficient C_r	Water Depth d (m)
Emerged	0.09	1.5	1.6	1.02	0.3
Submerged	0.1	2			0.4

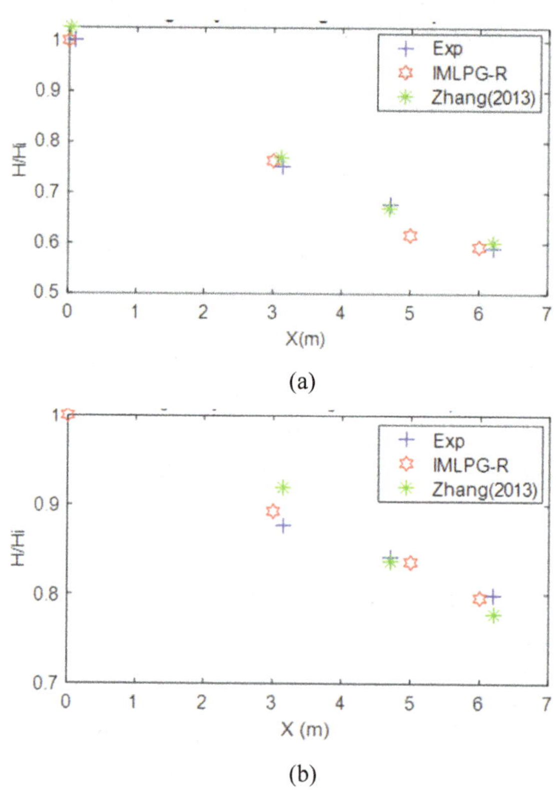

Fig. 8. Comparison of Experimental results (Augustin *et al.*, 2003), Numerical results (Zhang *et al.*, 2013) and IMLPG_R for (a) Emerged (b) Submerged cylindrical vegetation (Divya and Sriram, 2020).

In the numerical simulation, the computation domain length of 30.5m is considered with two different conditions, as in Table 2. The particle spacing is taken as 20 particles across the water depth and time step of 0.01s is considered. The results are compared with experiments and available numerical results, as shown in Fig. 8. A good agreement is obtained for the macroscopic model. Zhang *et al.* (2013) have used a numerical model based on the RANS equation with k-ε turbulence and solved using the finite difference method.

12.1. *Effect of Cd in cylindrical strips*

The coefficient of drag 'C_d' is a tricky parameter in the computation as the C_d value should represent the group effect of the vegetation and not the single stem. In order to understand the effect of C_d, a numerical investigation is carried out in Divya and Sriram (2020). The cylindrical vegetation cases reported in Table 2, for emerged is reproduced here, and the C_d values are changed in the computation. The results are shown in Fig. 9.

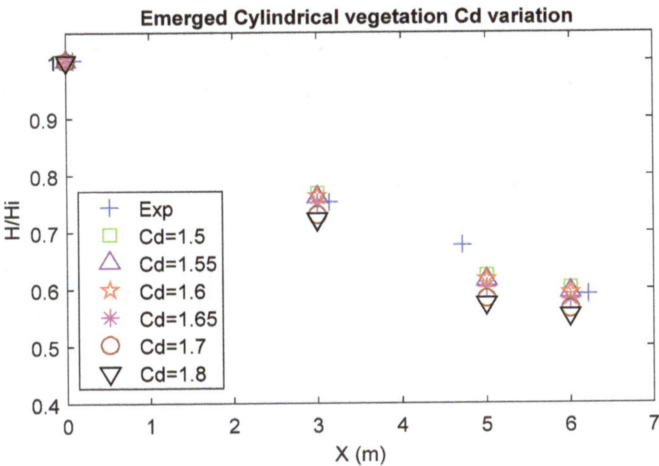

Fig. 9. Variation of result based on Cd value for Emerged (Divya and Sriram, 2020).

It can be seen from the Fig. 10(a), the variations in the wave attenuation are larger than the experimental measurements, even with slight variations in C_d. The C_d value is sensitive and plays a major role.

13. Comparison of the present model with microscopic 3D model

The developed model based on a macroscopic model where the vegetation region is assumed by semi-empirical equations, and the vegetation as such is not modeled. In order to see how the present model variation from the microscopic model, the results from Maza *et al.* (2015) are used. Maza *et al.* (2015) has developed a three-dimensional numerical approach based on IHFOAM to study the interaction of Tsunami waves with mangrove forest which they have represented for rigid cylinders. A direct simulation of vegetation region considering the vegetation geometry is carried out, which is similar to the microscopic model and also momentum damping using the drag force is simulated similar to the macroscopic model. An experiment is conducted for solitary waves and results are validated. To see the performance of the present IMLPG-R model developed based macroscopic level (Divya and Sriram, 2020), a numerical tank of length 20m and water depth of 0.15m is taken in which the vegetation region is started from 13m and ends at 14.635m. Vegetation height is taken as 0.3m with a diameter of 0.01m, having C_m value as 1. The C_d value is taken as 2 for wave height of 0.02m. The wave height is obtained at 10m, 13m, 13.545m, 14.09m, 14.635m and 14.935m. The wave height attenuation along the vegetation is compared with the experimental and numerical results from Maza *et al.* (2015) and given in Fig. 10.

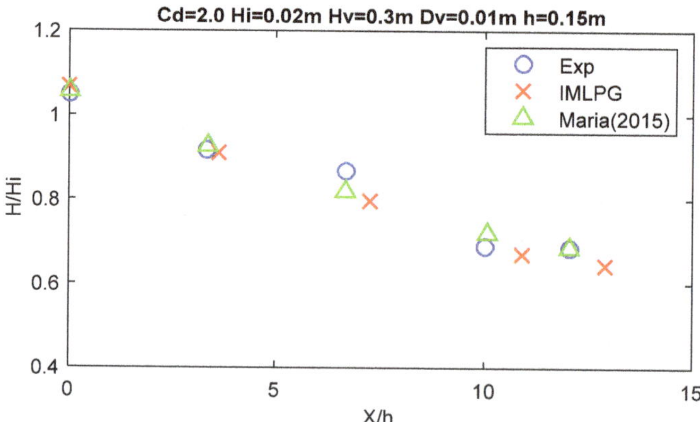

Fig. 10. Comparison of IMLPG_R results with Experiment and 3D numerical results from Maria (2015) for Cd=2.0.

It is evident from the above figure that the numerical results from IMLPG_R based on the macroscopic model are almost similar to that of Maza *et al.* (2015) results, which is based on the Eulerian framework and microscopic model. It should be noted that the present model is two dimensional but the model incorporated by Maza *et al.* (2015) was three dimensional.

14. Summary

In this chapter, existing numerical approaches for modelling porous and vegetation structure are reviewed. The combined modelling of the porous and vegetation structure in the unified governing equations are provided. So far these models are used for either modelling vegetation or porous structure (with multiple layer), however combined modelling of porous structure and vegetation together were not attempted in the literature. This is easy to incorporate and further investigation should be carried out in future. In order to circumvent the numerical instability at the interface due to sudden change in the porosity and resistance coefficient terms, proper interpolation scheme need to be adopted.

The macroscopic models reduces the computational effort significantly and the model results are equally good as the DNS model. Coastal defense structure focuses on the wave energy reduction for which a macroscopic model serves a preliminary estimations and vegetation movement is of least priority. However, the value of Cd or other coefficients used in the model has a high sensitivity thus the value must be accurate enough for reliable results.

Acknowledgement

This work is supported by the Department of Science & Technology, India Grant No. DST/CCP/CoE/141/2018C under SPLICE – Climate change Programme.

References

1. Akbari, H., M. Montazeri Namin (2013) Moving particle method for modelling wave interaction with porous structures, *Coastal Engineering*, 74, 59–73.
2. Brinkman, H.C., (1947) A calculation of the viscous force exerted by a flowing fluid on a dense swarm of particles. *Applied Scientific Research A1*, 27–34.
3. Carman, P.C., (1937) Fluid flow through granular beds, *Trans.Inst.Chem.Eng.*, 15, 150–66.
4. Cheng, Y.Z., Jiang, C.B., and Wang, Y.Y., (2009) A coupled numerical model of wave interaction with porous medium. *Ocean Engineering*, 36, 952–959.
5. Chorin, A.J. (1967) The numerical solution of the Navier-Stokes equations for an incompressible fluid, *Bull. Am. Math. Soc.*, 73, 928–931
6. Darcy, H., (1856) Les Fontaines Publiques de la ville de Dijon, Dalmont, Paris.
7. Engelund, F.A., (1953) On the laminar and turbulent flows of ground water through homogenous sand, *Danish academy of technical sciences, Denmark.*
8. Ergun, S., (1952) Fluid flow through packed columns, *Chem. Eng. Prog.*, 48, 89–94.
9. Fair, G.M., Geyer, J.C. and Okum, D.A., (1968) Water purification and wastewater treatment and disposal, *Vol 2, Waste water management*, Wiley, New York.
10. Forchheimer, P., (1901) Wasserbewegung durch Boden. Zeitschrift des Vereins Deutscher Ingenieure 45, 1782–1788.
11. Huang, C.J., Shen, M.L., and Chang, H.H., (2011) Propagation of a solitary wave over rigid porous beds. *Ocean Engineering* 35, 1194–1202.
12. Lin P., and Liu, P.L.-F., (1998) A numerical study of breaking waves in the surf zone. *Journal of Fluid Mechanics* 359, 239–264.
13. Liu G.R., Gu Y.T., (2005) An introduction to meshfree methods and their programming. CRC press.
14. Liu, P.L.-F., Lin, P., Chang, K.A., and Sakakiyama, T., (1999) Numerical modeling of wave interaction with porous structures. *Journal of Waterway, Port, Coastal and Ocean Engineering* 125, 322–330.
15. Karunarathna, S.A.S and Lin, P., (2006) Numerical Simulation of wave damping over porous sea beds. *Journal of Coastal Engineering* 53, 845–855.
16. Lin, P. and Liu, H.W., (2007) Scattering and trapping of wave energy by a submerged truncated paraboloidal shoal, *J. Waterways. Port Coast. Ocean Eng. ASCE*, 133(2), 94–103.
17. Ma, Q.W., (2005a) Meshless Local Petrov Galerkin method for two dimensional non-linear water wave problems. *Journal of Computational Physics* 205, 611–625.
18. Ma, Q.W., (2005b), MLPG Method Based on Rankine Source Solution for Simulating Nonlinear Water Waves, *CMES: Computer Modelling in Engineering & Sciences*, Vol. 9, No. 2, pp. 193–209.
19. Ma, Q.W., (2008) A new meshless interpolation scheme for MLPG-R method, *CMES*, vol 23(2), 75–89.

20. Ma, Q.W., and Zhou, J., (2008) Numerical Implementation of solid boundary conditions in Meshless methods, *Proceedings of eighteenth international offshore and polar engineering conference*, Vancouver, BC, Canada, July 6–11.

21. Sriram, V. and Ma, Q.W., (2012). Improved MLPG_R for simulating 2D interaction between violent wave and elastic structures, *Computational Physics*, 231, 7650–7670.

22. Polubarinova-Kochina, P.Y., (1962) Theory of ground water movement (translated from the Russian by J.M.R De Wiest). Princeton University Press.

23. Ren, B., Wen, H., Dong, P., and Wang, Y., (2014) Numerical simulation of wave interaction with porous structures using an improved smoothed particle hydrodynamic method, *Coastal Engineering*. 88, 88–100.

24. Ren, B., Wen, H., Dong, P., and Wang, Y., (2016) Improved SPH for wave motions and turbulent flows through porous media. *Coastal Engineering*, 107, 14–27.

25. Shao, S., (2010) Incompressible SPH flow model for wave interactions with porous media, *Coastal Eng.* 57 (3), 304–316.

26. Van Gent, M.R.A., (1995) Wave interaction with permeable coastal structures, Ph.D. thesis, Delft University of technology, ISBN 90-407-1182-8, Delft University Press, Delft, The Netherlands.

27. Ward, J.C., (1964) Turbulent flow in porous media. Proceedings, *Journal of Hydraulics Division*, ASCE, 90 (HYS). 1–12.

28. Sollitt, C.K. and Cross, R.H., (1972) Wave transmission through permeable breakwaters. *Proc.13th Coastal Eng. Conf.*, ASCE, Vol. III, pp. 1827–1846.

29. Dean, R.G., Dalrymple, R.A., (1984.) *Water Wave Mechanics For Engineers and Scientists*, Prentice-Hall, p. 353.

30. Su, X., Lin, P., (2005). A hydrodynamic study on flow motion with vegetation, *Mod. Phys. Lett. B* 19 (28–29) 1659–1662.

31. Suzuki, T., Hu, Z., Kumada, K., Phan, L.K., Zijlema, M., (2012). Non-hydrostatic modelling of drag, inertia, and porous effects in wave propagation over dense vegetation fields, *Coastal Eng.* 149 49–64.

32. Yao, Y., Tang, Z., Jiang, C., He, W., Hiu, Z., (2018) Boussinesq modelling of solitary wave run up reduction by emergent vegetation on a sloping beach, *J. Hydroenviron. Res.* 19 78–87.

33. Maza, M., Lara, J.L., Losada, I.L., (2015). Tsunami wave interaction with mangrove forests: a 3-D numerical approach, *Coastal Eng.* 98 33–54.

34. Maza, M., Lara, J.L., Losada, I.L., (2016). Solitary wave attenuation by vegetation patches, *Adv. Water Resour.* 98 159–172.

35. Divya, R., Sriram, V., (2017). Wave-porous structure interaction modelling using Improved meshless local Petrov-Galerkin method, *Appl. Ocean Res.* 67 291–305.

36. Divya, R., Sriram, V., (2020). Wave-vegetation interaction using Improved meshless local Petrov-Galerkin method, *Appl. Ocean Res.* 101 102116.

37. Agarwal, S., Sriram, V., Murali, K., (2019). Modelling wave interaction with porous structures using Boussinesq equations, *Proceedings of the Fourth International*

Conference in Ocean Engineering, ICOE2018, Springer, Singapore (2019), pp. 573–583.

38. Agarwal, S., Sriram, V., Liu, P.-F., Murali, K. (2022). Waves in waterways generated by moving pressure field in Boussinesq equations using unstructured finite element model *Ocean Eng.*, 262 (2022), Article 112202,

39. Asano, T., Deguchi, H., Kobayashi, N., (1992) Interaction between water waves and vegetation, *Proceedings of the 23rd international conference on coastal engineering*, Venice, Italy, pp. 2710–2723 Chapter 207.

40. Marsooli, R., Wu, W., (2014) Numerical investigation of wave attenuation by vegetation using a 3D RANS model, *Adv. Water Resour.* 74 245–257.

41. Augustin, L.N., Irish, J.L., Lynett, P., (2009) Laboratory and numerical studies of wave damping by emergent and near emergent wetland vegetation, *Coastal Eng.* 56(3) 332–340.

42. Zhang, M., Hao, Z., Zhang, Y., Wu, W., (2013) Numerical simulation of solitary and random wave propagation through vegetation based on the VOF method, *ActaOceanol. Sin.* 32(7) 38–46.

Chapter 8

Turbulence Structure due to Wave-Vegetation Interaction

Zhihua Xie

School of Engineering, Cardiff University,
The Parade, Cardiff, CF24 3AA, UK
zxie@cardiff.ac.uk

Thorsten Stoesser

Department of Civil, Environmental and Geomatic Engineering,
University College London
Gower Street, London, WC1E 6BP, UK
t.stoesser@ucl.ac.uk

In this chapter methods for simulating wave-vegetation (WVI) interaction are reviewed, with an emphasis on approaches that are resolving numerically the large and small-scale dynamics in the near-field of aquatic plants or surrogates thereof. The fluid flow around such (fixed or moving) is complex, turbulent and highly three-dimensional and local flow instabilities result in flow separation, turbulence production in the form of coherent flow structures and energy dissipation. The accurate representation of these processes in numerical models that are aiming at gaining fundamental insights into the mechanisms of wave-vegetation interaction and/or at deriving physics-based subgrid scale models for large-scale applications of WVI. First, the governing equations suitable for near-field wave-vegetation large-eddy simulations are introduced together with selected numerical approaches to handle the complex physics. Secondly, examples of eddy-and plant-resolving simulations of flow through vegetation surrogates are presented.

1. Introduction

The impact of climate change and sea level rise triggers weather extremes that cause significant damage to coastal communities. Protecting our coastal assets and population from the effects of climate change will be one of the great engineering, logistical, socio-economic and political challenges of this century. Coastal vegetation, such as trees, mangroves, salt marshes, and seagrasses can attenuate wave energy and mitigate coastal flooding.[1] Thus, better understanding of wave-vegetation interaction (WVI) is essential to the future design of nature-based coastal defences and/or natural disaster mitigation strategies.

A number of physical experiments have been carried out to investigate the effect of vegetation on wave attenuation[2-6] and wave-current interaction.[7-9] Some of the experiments focus on a single solitary propagation,[3,6] and there are also studies on regular and irregular waves.[2,4,5,10,11] Regarding the effect of vegetation, experiments were conducted investigating plant stem, root and canopy effects,[4] plant morphology, flexibility, and shoot density,[5,12,13] suspend vegetation,[11] or large-scale 3D effects, respectively.[7] A detailed review of flow and transport process within aquatic vegetation can be found in.[14]

In addition, there are also several theoretical studies aiming at revealing wave dissipation mechanism and drag forces of vegetation. Empirical models have been developed originally for emergent rigid cylinders,[15] and then further extended for submerged vegetation,[16] vegetation under nonbreaking and breaking waves,[17] wave-current interaction,[18] long waves,[19] a general framework for emergent, submerged and floating vegetation,[20] and viscous boundary layers and vertical shear stress distributions.[21]

With the development of mathematical models and numerical methods, computational approaches have been becoming popular tools to unveil the energy dissipation mechanism in WVI and provide more detailed information of the flow around/above vegetation. Different wave hydrodynamic models have been developed, either with a phase-averaging approach, for example the spectral wave model SWAN[22] and mild slope equation,[23] or the more commonly used phase-resolving approach. In phase-resolving models, there are depth-integrated numerical models based on the Boussinesq[24,25] and shallow water equations with hydrostatic pressure[26,27] or non-hydrostatic pressure as used in the SWASH[28] and XBeach[29] codes; there are also non-hydrostatic 3D wave models[30-33] and σ-coordinate 3D model.[34] However, breaking waves cannot be considered in these depth-integrated and non-hydrostatic models. In order to overcome this limitation, numerical models have been developed based on the full Navier-Stokes equations with interface calculation methods to predict the air-water interface. Some examples include 2D[35,36] and 3D[3,37] Reynolds-averaged Navier–Stokes (RANS) model, or the eddy-resolving large-eddy simulation

(LES) model,[38] respectively.

For wave hydrodynamic models based on the Navier-Stokes equations, three approaches are often used to take into account vegetation: (a) via a (subgrid-scale) drag force in the momentum equation e.g.;[35] (b) via energy dissipation mechanism, e.g.;[39] or (c) via a porous media[40] approach. Most numerical studies are mesh-based method, such as for rigid vegetation,[32] quasi-flexible vegetation,[41] flexible vegetation,[36,42,43] and wave-current-vegetation interaction.[35,44] Recently, there have been Meshless Local Petrov Galerkin (MLPG) methods[45] and Smooth Particle Hydrodynamics (SPH) methods[46] developed which could be applied to WVI. In the following sections various aspects of numerical WVI are considered.

2. Numerical Methods for Simulating and Modeling Wave-Vegetation Interaction

With the rapid increase of computational power and development of powerful computational fluid dynamics (CFD) methods, several numerical techniques have been proposed to study wave-vegetation interaction (WVI). The accurate simulation of WVI is a complex, multi-dimensional process, which takes place at a large range of spatial and temporal scales. A numerical simulation approach requires a number of key ingredients, i.e. (i) an accurate fluid solver which, ideally, resolves all relevant scales so that governing physical processes are captured correctly, (ii) adequate wave generation and dissipation techniques to provide physically correct boundary conditions, (iii) an accurate structural solver which calculates the movement of flexible bodies when subjected to hydrodynamic forces and (iv) a fluid-structure interaction method which transfers forces between vegetation and fluid in a realistic way.

2.1. *Fluid solver*

Most fluid solvers for complex WVI simulations are based on the Navier-Stokes equations for incompressible flows which read:

$$\boldsymbol{\nabla} \cdot \bar{u} = 0, \tag{1}$$

$$\frac{\partial(\rho \bar{u})}{\partial t} + \boldsymbol{\nabla} \cdot (\rho \bar{u} \otimes \bar{u}) = -\boldsymbol{\nabla}\bar{p} + \boldsymbol{\nabla} \cdot [\mu(\boldsymbol{\nabla}\bar{u} + \boldsymbol{\nabla}^T \bar{u})] + \rho g + F + \boldsymbol{\nabla}\tau^{\text{tur}}. \tag{2}$$

where the overbar $\bar{\cdot}$ denotes either Reynolds-averaged or spatially-filtered quantities over the grid in Cartesian coordinates (x, y, z), $\bar{u} = (\bar{u}, \bar{v}, \bar{w})$ and

$\bar{p}$ are the velocity vector and pressure. t is the time and g is the gravitational acceleration vector. ρ and μ are the density and dynamic viscosity of the fluid, which are constant in a single-phase flow model, while they are phase-dependent in two-phase flow models. The term F is a body force, an external force acting on the fluid flow due to vegetation as discussed in 2.5 and 2.7. The term τ^{tur} are either Reynolds stresses in a RANS modelling or the subgrid-scale (SGS) stress tensor in LES framework as discussed in 2.6. In WVI, inherently, three phases are present, (i) water, (ii) air and (iii) solid and the fluid solver needs to be able to deal with all phases (arguably the effect of the air phase on the process of WVI is negligible) as well as the interfaces between phases (again, arguably, the air-vegetation interface can be neglected). The water-air interface is, in 3D RANS simulations or LES, usually calculated by interface tracking or capturing methods, respectively, for instance by the volume-of-fluid, level-set, phase field, or particle methods. If the vegetation is resolved explicitly by the grid, the water-vegetation interface is treated with a no-slip condition. Many times only the effect of the vegetation drag on the fluid, e,g, in large-scale wave-attenuation studies, needs to be calculated and for these simulations a subgrid-scale vegetation closure model is employed to estimate the momentum loss. On other solid boundaries, no slip or wall-function boundary conditions are used.

2.2. *Wave generation and dissipation*

WVI is different from flow through submerged or emergent vegetation in open-channels,[47,48] where waves are absent: the flow around vegetational elements is not constant or uni-directional, respectively. A so-called numerical wave tank has to be incorporated into the fluid solver, i.e. waves are generated at the inlet, propagate inside the fluid domain and are damped/dissipated at the domain outlet. For wave generation, three main approaches can be used: (a) specifying the wave elevation and particle velocities at the inlet as Dirichlet boundary conditions with or without relaxing zones;[49,50] (b) wave generation inside the domain as internal source terms;[51] or (c) reproduction of the exact motion of a wavemaker to generate the waves as is done in laboratory experiments.[46,52,53]

For wave dissipation, either numerical damping zones or numerical beach can be added towards the end of the computational domain to dissipation the wave energy in order to reduce the wave reflection from the outlet.[49,54,55]

2.3. *Structural solver*

Depending on the required level of complexity of the WVI simulation, different structural solvers can be employed to compute the response of the vegetation to the flow/wave. The simplest approach is to assume the vegetation as rigid body, hence no structural solver is needed, the vegetation's flexural rigidity (often denoted EI, where E is Young's modulus and I is the second moment of inertia) is infinite and the vegetation does not bend or move in response to the hydrodynamic load. With this approach, e.g.,[34,35,37,56] only the force exerted by the vegetation on the fluid is required and thus it is usually referred to as one way coupling. In reality, vegetation is a flexible structure and depending on its EI value its motion may become important to the WVI, for instance by altering its flow resistance while under load (e.g. seagrass, a highly flexible material adapts to the local hydrodynamics and thereby exhibits significant movement). Such behavior is referred to two-way coupling, i.e. the hydrodynamic force moves the vegetation and in turn the presence of the structure or its force alters the local flow field. Several structural solvers have been developed to model the deformation of the vegetation,[33,36,41–43,57,58] by considering it as a 1D, 2D, or 3D objects.

2.4. *Numerical fluid-structure interaction*

There are mainly two different methods to accomplish numerically fluid-structure interaction. In boundary-fitted methods, the numerical mesh conforms to the geometry of the structure and as the structure moves the mesh is distorted. A boundary-fitted mesh is straightforward for very complex vegetation geometries when the structure is stationary. However, when the vegetation is moving and deforming, a body-fitted mesh can become cumbersome and requires re-meshing at every time step. In certain situations, for instance when highly-flexible seagrass is forced flat on the bed, a body-fitted mesh is not even possible anymore. The alternative, Cartesian grid methods, has become popular with the advent of immersed boundary,[59] ghost cell[60] or cut-cell[61] methods which allow solving fluid flow or wave propagation on a fixed Cartesian grid while structures or in this case vegetation elements can take any shape or allow for any movement inside the Cartesian grid.[62,63] These methods have the additional benefit of super simple and fast mesh generation and easy data structure management.[64] The structural part of the "immersed body" can be achieved easily using finite element (FEM) or a Lagrangian model flexible vegetation problems.

2.5. *2D and 3D WVI*

Due to the complexity of the physical process of WVI and the generally high computational cost, most previous WVI studies have been restricted to 2D simulations, in which a 2D wave model is used.[22,26] For rigid vegetation, a so-called canopy zone model can be used in which the canopy density is varied to represent the effect of vegetation. A force balance equation is considered for rigid and stationary objects, accounting for the drag, added mass, and virtual buoyancy[65] and added as a sink term in the momentum equation. In such subgrid vegetation closure models, the drag coefficient is an important model parameter, which may be a constant (e.g. for uni-directional flows) or vary as a function of the Keulegan–Carpenter (KC) number under wave conditions. For flexible vegetation, a finite element method (FEM) model is normally used with a 1D line model (a slender elastic rod theory)[36] to represent the objects and then coupled with the 2D hydrodynamic model. Such 2D approaches allow for fast computations of large-scale applications but rely on empirical input and a number of assumptions.[57]

For the simulation of more realistic, smaller-scale scenarios of wave-vegetation interaction, 3D simulations should be considered. In most simulations to date, the vegetation is mimicked as 3D circular cylinders arranged in arrays and individual elements are resolved explicitly by the grid, instead of being modelled as a momentum sink as in 2D simulations.[38] At a significantly higher computational expense, the details of fluid-structure interaction is obtained from such 3D simulations and provides physical insights into the vegetation-scale, unsteady hydrodynamic features, including vortex shedding from the vegetation, dynamic lift and drag forces acting on the vegetation, turbulence statistics, and the fate of 3D coherent structures in the wake. It is worth noting that most 3D WVI simulations are based on simple cylinder geometries, high-fidelity numerical WVI research on complex geometries to replicate the shape of the real vegetation remain yet to be explored.

It is challenging to simulate the complex 3D fluid-structure interaction of WVI for flexible vegetation, which requires two-way coupling between the fluid solver and the structural solver. As mentioned above, complex structural solvers are usually based on the FEM allowing for detailed 3D assessment of the movement or deformation of solids, respectively. This adds additional complexity and computational cost in WVI and hence, to date, most studies assumed slender objects, in which a (simplified) beam theory (1D line model) is employed to account for the flexibility of vegetation in 3D simulations of WVI.[38,46]

2.6. *RANS vs LES*

In general, wave-vegetation interaction is a turbulent hydrodynamic process and therefore the effect of turbulence on the mean and instantaneous flows requires accounting for unless all scales of turbulence are resolved, which is, for practical applications not feasible due to the exorbitant computational demands of a so-called Direct Numerical Simulation (DNS). In many previous studies of WVI, the effects of turbulence have been neglected as these were deemed negligible.[27,29,57] Recent endeavours have employed the Reynolds-averaged Navier–Stokes (RANS) equations, in which all of the unsteadiness due to turbulence is averaged out and the effects of turbulence are modelled by a turbulence model.[30,35–37] Thus, RANS models cannot provide instantaneous turbulent flow characteristics. The increase in computer power has led to the development of more powerful but more computationally demanding methods, such as the method of large-eddy simulation (LES),[47,66,67] in which large-scale turbulence is resolved by the numerical method while the effect of small scale turbulence on the large scales is modelled. For instance,[68] performed single-phase LES to study wave impact on a slender cylinder while[54,62,69] carried out two-phase flow LES for three-dimensional wave-structure interactions and demonstrated the feasibility of this approach to study WVI.[38] used the OpenFOAM code with the LES approach to investigate both unidirectional and wave flows through vegetation. In comparison to RANS-based methods, the method of LES offers increased accuracy, however the computational cost of LES is significantly greater than the RANS counterpart. On the other hand, With the ever-increasing computational power LES is already, and will be increasingly in the future, applied for practical problems where the Reynolds number is high and the computational domain is large.

2.7. *Vegetation closure models*

When a large-scale RANS approach is employed, the form drag due to vegetation is accounted for through subgrid forces that are added to the momentum and turbulence model equations, which may be referred to as vegetation closure model.[70] Figure 1 provides a good summary of 2D and 3D numerical simulations of unidirectional flow through vegetation. The drag coefficient is an empirical parameter and has been kept constant in some work, most likely due to the lack of detailed knowledge of it. In other simulations the drag coefficient is a function of Reynolds number, vegetation density, or the KC number for waves, respectively. This table demonstrates the variability of this coefficient for fairly simple, cylindrical, rigid vegetal elements which in turn suggests an even wider bandwidth of values (and uncertainty) for non-rigid vegetation of complex geometrical shape.

Study	Model Approach/Turbulence Closure	Drag Force Closure (Drag Coefficients)	Parameter (Coefficient) Treatment
Tsujimoto and Shimizu [1994]	2D RANS k-ε model	Drag force terms to momentum equations (C_{db}, C_{dz})	no values provided for C_{db}, C_{dz}
		Drag force terms to turbulence energy equations (C_{fk}, $C_{f\varepsilon}$)	$C_{fk} = 1.0$, $C_{f\varepsilon} = 1.3$
Wu et al. [2005]	2D RANS k-ε model	Drag force terms to mom. equations (C_d)	$C_d = 1.0 - 4.0$
		Drag force terms to turbulence energy equations (C_{vk}, $C_{\varepsilon 3}$)	$C_{vk} = 1.0$, $C_{\varepsilon 3} = 1.33$
Naot et al. [1996]	RANS k-ε model	Drag force terms to mom. equations (C_D)	$C_D = (10^3/Re_D)^{0.25}$ for $Re_D = \leq 10^3$ or $C_D = $ minimum of
		Drag force terms to turbulence energy equations (C_D)	$0.976 + [(10^{-3}Re_D-2)/20.5]^2$ or 1.15 for $10^3 < Re_D < 4 \times 10^4$ [*Schlichting*, 1962]
Lopez and Garcia [2001]	RANS k-ε and k-ω model	Drag force terms to mom. equations (C_D)	$C_D = 1.13$ [*Dunn et al.*, 1996]
		Drag force terms to turbulence energy equations (C_{fk}, $C_{f\varepsilon}$)	$C_{fk} = 1.0$, $C_{f\varepsilon} = 1.33$
Fischer-Antze et al. [2001]	RANS k-ε model	Drag force terms to mom. equations (C_D)	$C_D = $ is set as 1.0
Neary [2003]	RANS k-ω model	Drag force terms to mom. equations(C_D)	$C_D = 1.13$ [*Lopez and Garcia*, 2001] for case 1 and $C_D = 1.0 - 1.5$ for case 2
		Drag force terms to turbulence energy equations (C_{fk}, $C_{f\omega}$)	$C_{fk} = 1.0$ [*Lopez and Garcia*, 2001], $C_{f\omega} = 1.5$ for case 1 and $C_{fk} = 0.05$, $C_{f\omega} = 0.16$ for case 2
Nicholas and McLelland [2004]	RANS k-ε model	Drag force terms to mom. equations (C_D)	C_D is set as 1.0
Choi and Kang [2004]	Reynolds stress model	Drag force terms to mom. equations (C_D)	$C_D = C_{DA} f(z, h_p)$, $C_{DA} = 1.13$ [*Dunn et al.*, 1996] with $z = $ depth and $h_p = $ vegetation height
Defina and Bixio [2005]	RANS k-ε model	Drag force terms to mom. equations (C_D)	$C_D = 1.0 - 3.0$
		Drag force terms to turbulence energy equations (C_{fk}, $C_{f\varepsilon}$)	$C_{fk} = 1.0$, $C_{f\varepsilon} = 1.33$ [*Lopez and Garcia*, 2001] and $C_{fk} = 0.07$, $C_{f\varepsilon} = 0.16$ [*Shimizu and Tsujimoto*, 1994]
Nadaoka and Yagi [1998]	2D LES	Drag force terms to 2D mom. equations (C_b)	$C_b = 2.49$ for *Tsujimoto and Kitamura*'s [1992] case and $C_b = 2.42$ for *Ikeda et al.*'s [1994] case
		A drag force term to energy transport equation (C_b)	
Cui and Neary [2002]	LES	Drag force terms to mom. equations (C_D)	$C_D = 1.6$
Stoesser et al. [2009, 2010]	LES	Explicit resolution of circular cylinders imposing the no-slip boundary condition	

Fig. 1. Formulations to Characterize vegetation drag by different researchers. (The picture is taken from[70]).

In highly-resolved LES, a vegetation closure model is not necessarily needed as the vegetation geometry could be resolved explicitly by the grid and a direct application of no-slip boundary conditions on the vegetation. With such treatment, the local instantaneous flow around vegetation is properly resolved and vegetation drag is implicitly accounted for in the most accurate way. The drawback of such an approach is its enormous computational expense and such LES are limited to small-scale laboratory-size studies or canonical flows from which drag coefficients of more realistic vegetation could be determined for large-scale RANS models.

3. Recent advances in WVI

In this section a series of numerical simulations of wave-vegetation interaction are presented, ranging from wave-single-rigid-cylinder to wave-flexible-vegetation-array interaction with an increasing degree of complexity.

3.1. *Breaking wave-cylinder interaction*

It is worth noting that previous WVI studies focused on a single wave or a train of regular and irregular waves without wave breaking. When there is a breaking wave interacting with a cylinder, the flow physics and associated turbulence structures are much more complex. Figure 2 visualises Stokes wave breaking in a periodic domain. The simulation is initialised with a modified potential solution of water waves, the third-order Stokes wave solution for the air-water interface and the velocity potential in the water. The velocity in the air is initialised as zero as little is known about the air movement and due to its low density air movement is deemed irrelevant here. The computational domain spans $2L \times L \times 0.5L$ (where L is the wave length) in the streamwise, vertical, and spanwise direction, respectively. Periodic boundary conditions are used in both streamwise and spanwise directions. The Cartesian grid multiphase flow solver Xdolphin3D[63,69] is employed and it is based on the large-eddy simulation (LES) approach with the dynamic Smagorinsky subgrid-scale model. The governing equations are discretised using the finite volume method on a staggered Cartesian grid. A Cartesian cut-cell method[61] is developed to deal with the complex topography. The PISO algorithm couples the pressure to the velocity and a second-order backward Euler method approximates the time derivative. The model has been further developed to capture the air-water interface by a geometric PLIC-VOF (Piece-wise Linear Interface Calculation Volume-of-Fluid) method, and the continuum surface force model is implemented for surface tension effects. It can be seen that both the plunging jet, jet-splash cycles and the air entrainment are well captured with Xdolphin3D.

Fig. 2.   Snapshot of air entrainment under breaking waves using the Xdolphin3D code.[63]

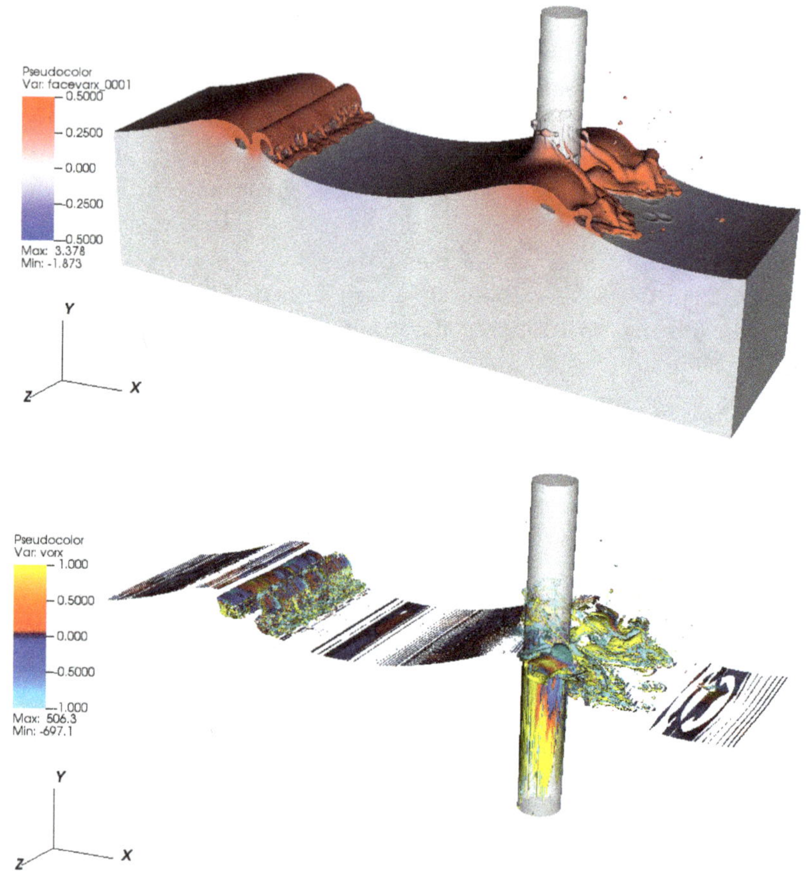

Fig. 3.   Snapshot for the air-water interface colored by the streamwise velocity (top) and vortical structures colored by the streamwise vorticity (bottom) during wave breaking-cylinder interaction using the Xdolphin3D code.[63]

Figure 3 shows an example of breaking-wave-vegetation interaction, in which the vegetation element is represented as a single circular cylinder and the vortical structures are identified by the λ_2 method. The computational setup is similar to the case above and a cylinder is located at the wave breaking point for the first wave whereas strong wave impact is expected. It can be seen that without vegetation, the splash-up is generated when the plunging jet hits the water surface ahead and a two-dimensional air cavity is enclosed under the main jet. When there is a cylinder during wave breaking,

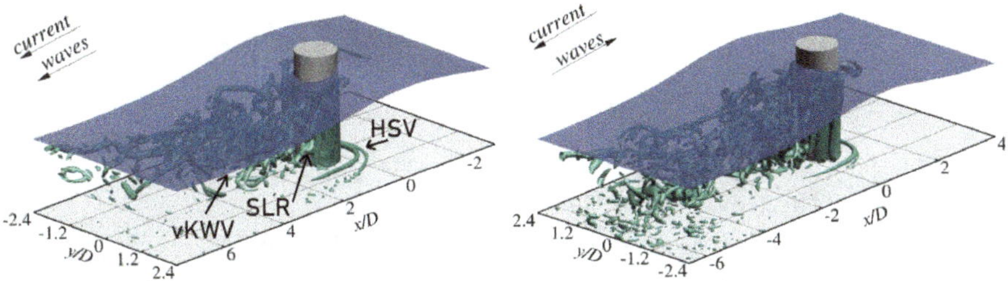

Fig. 4.   Turbulence structures in wave-current-vegetation interaction visualised by iso-surfaces of the Q-criterion for a following current (left) and an opposing current (right).

the plunging jet slams the cylinder and moves upwards and wraps around the cylinder.   More vortical structures are generated during the impact resulting in a fully three-dimensional process, which is responsible for more energy dissipation and vortex shedding. Small-scale bubbles and droplets are also generated during the wave-breaking process.

3.2. *Wave-current-cylinder interaction*

The visualisation of a wave-current-vegetation interaction is presented in Figure 4 in which a focused wave with a following current (left) and one with an opposing current (right) interacts with a rigid circular cylinder. The result is obtained via a high-resolution large-eddy simulations which were performed with the open-source code Hydro3D-NWT,[54] a Navier-Stokes equations solver that is based on Cartesian grids and which uses the WALE subgrid-scale model. Hydro3D-NWT employs the level set (LSM) and immersed boundary (IB) methods to handle accurately the interfaces between water and air (LSM) and water and solid (IB). To generate waves in Hydro3D-NWT, Dirichlet boundary conditions based on kinematics obtained from analytical solutions matching the observations from an experiment. At the inlet a shear flow superimposed with the waves' velocities are prescribed as:

$$u_i(z) = u_{iw}(z) + u_{is}(z) \tag{3}$$

where $u_i(z)$ is the total velocity vector set at the inlet boundary as a result of the wave induced velocity $u_{iw}(z)$ and current's velocity $u_{is}(z)$ at a distance z from the floor bed. To account for the variation of water depth due to water surface oscillations, 3 is modified as:

                          Z. Xie and T. Stoesser

$$u_i(z_s) = u_{iw}(z_s) + u_{is}(z_s) \tag{4}$$

where the new variable z_s is defined as:

$$z_s = \frac{z}{n(t)} * d \tag{5}$$

where $n(t)$ is the elevation of the water surface at the wave-maker location and d is the still water level. In the vicinity of the outlet Hydro3D-NWT absorbs waves using the relaxation method.[49] This method employs a relaxation function $\Gamma(X)$ as follows:

$$\Gamma(X) = 1 - \frac{e^{X^R} - 1}{e - 1} \quad , \quad X = \frac{x - x_s}{x_e - x_s} = [0, 1] \tag{6}$$

where X is a non-dimensional measure of the relaxation zone, usually two wavelengths long $(x_e - x_s = 2L)$ and R is set to 3.5 as in.[49] To retain still water conditions the following equations are used:

$$\phi = (1 - \Gamma(X))\phi_{target} + \Gamma(X)\phi_{computed} \tag{7}$$

$$u_i = (1 - \Gamma(X))u_{i_{target}} + \Gamma(X)u_{i_{computed}} \tag{8}$$

Initially, the water surface is set to the still water level ($\phi_{target} = 0$ at $z = d$) based on eq. 7 and velocities to the desired values $u_{i_{target}}$ using eq. 8. In the case of waves and shear currents, the velocities at the outlet of the domain are set to the corresponding current velocity to ensure constant mass throughout the simulation, i.e. $u_{i_{targeted}} = u_{i_s}(z)$. The numerical grid employed for this simulation is depicted in Figure 5. It comprises approximately 26 million cells and uses a local-mesh refinement in the vicinity of the cylinder and the water surface which allows to resolve steep gradients of the wave-vegetation interaction.

In a first step measurement data from an experiment were used to assess the accuracy of the code to reproduce wave-current kinematics in the tank, without structure, for a following, an opposing and no current cases, all under the same wave conditions. The LES results in terms of wave elevations and water velocities under the crest and trough match very well the experimental data for all conditions.[71] Convergence and validation studies were carried out based on current-cylinder only simulations, followed by two LES of wave-current-cylinder interaction. Figure 4 depicts the turbulence structures that are expected to occur locally in the vicinity of the cylinder/vegetation element. Three of these turbulence structures are pointed

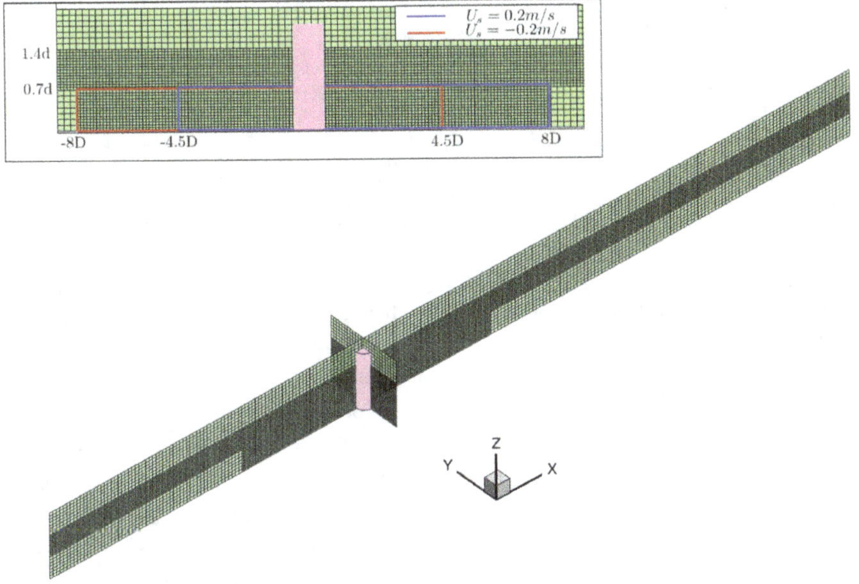

Fig. 5. The Cartesian grid with local mesh refinement employed for the wave-current-vegetation interaction LES.

out in the following current simulation and are horseshoe vortex, HSV, shear layer vortices, SLR and von Karman-type wake vortices. The shear current creates a boundary layer in the water column and this boundary layer rolls up as the flow approaches the upstream side of the cylinder, this vortex then wraps around the cylinder as the fluid passes and forms a vortex of the shape of a horseshoe. The shear layer at the sides of the cylinder similarly rolls up into a vortex, the axis of which is vertical and this vortex is shed periodically and simultaneously on both sides from the structure before it is convected downstream. The wake of the cylinder is characterised by a number of elongated vortices which are the remains of the SLR vortices plus some wake generated turbulence. The analogue structures of the opposing current case are quite similar in size and location, suggesting that the waves contribute less to the fluid-structure interaction than the current. However, the turbulence in the wake is clearly enhanced as the wave propagates upstream. Another way of visualising turbulent vortices in the fluid is by contours of the swirl strength in 2D planes. Such contours in a horizontal plane at half the still water depth are presented in Figure 6.

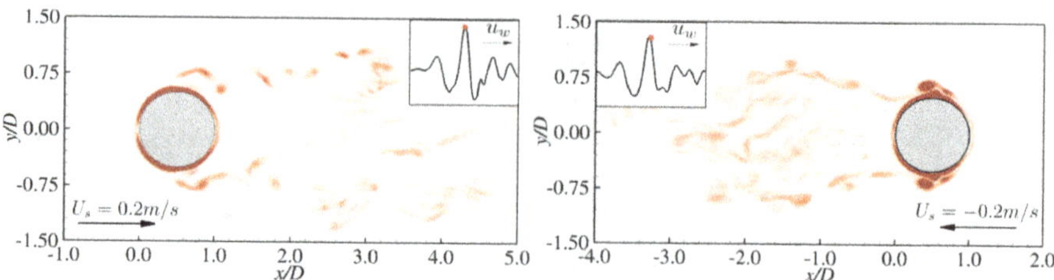

Fig. 6.   Contours of the swirl strength in horizontal planes at half the still water depth for wave-current-vegetation interaction for a following current (left) and an opposing current (right).

In the following current scenario, the vortices that are separating at the cylinder side are pushed downstream and due to the additional acceleration of the fluid due to the wake take a wider path around the structure, hence the wake is wider. The opposing current scenario is characterised by a decelerated, narrower wake and the flow separation of the SLR vorices is delayed, almost stalled at the instant depicted. A second SLR vortex is already formed while the previous one has barely moved downstream.

Such high-resolution simulations can shed light on a variety of physical phenomena and can contribute to a general understanding of WVI, albeit at high computational expense and at the moment limited to relatively simple geometries and low Reynolds numbers.

3.3. *Wave-cylinder-array interaction*

The in-house CFD code hydro3D based on the large eddy simulations method is employed to reproduce the experiments of a solitary wave interacting with various piercing cylinders to simulate the effect of vegetation in near coast environments. A solitary wave of wave height $H/d = 0.52$ is generated using Dirichlet boundary conditions based on[72] inside a 13 m long, 0.6 m wide and 0.6 m deep. The still water depth is $d = 0.25$ m and the array of cylinders is placed 6.15 m away from the wavemaker similar to the laboratory study. No-slip boundary conditions for the floor bed and cylinder surfaces are adopted where the side walls and top boundary are set to free-to-slip conditions. A numerical beach of 4 m long near the outlet of the domain is also used based on the relaxation method[49] to absorb incident waves.

Figure 7 visualizes the interaction of the water surface with the vegetation, coloured with the normalized elevation (n/H), at various time steps.

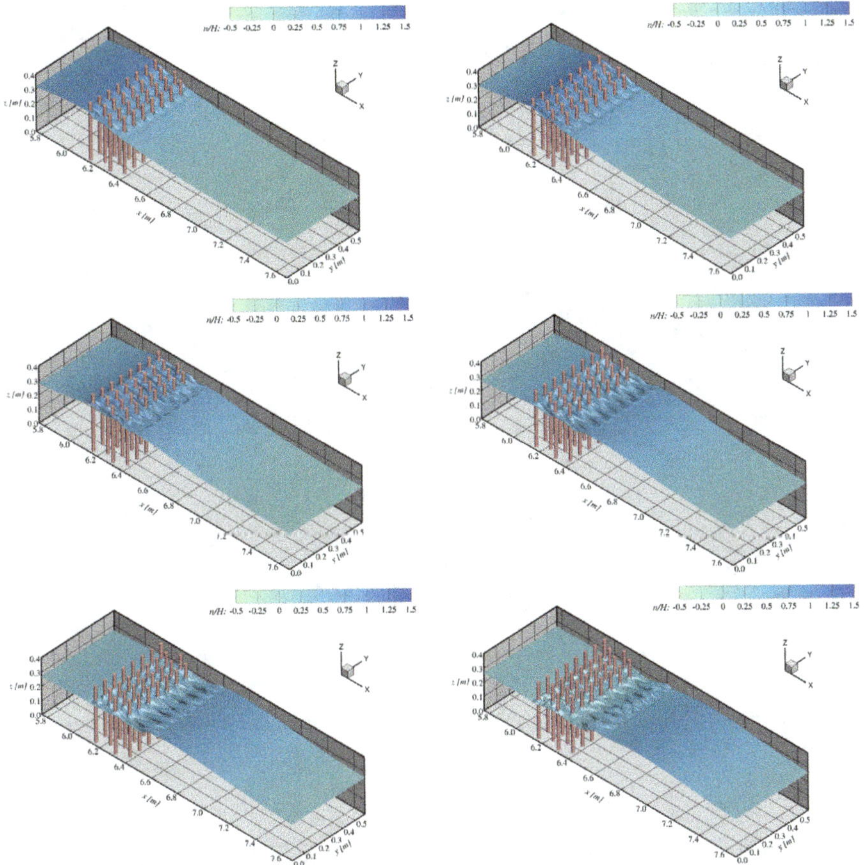

Fig. 7. Snapshots for the water surface profile during wave-cylinder-array interaction using the Hydro3D code.[54]

In Figure 1a ($t = 4.6$ sec) the wave has reached the array and the flow accelerates through the gaps between cylinders, forming small standing waves in the front faces of the vegetation and a shallower wake on the rear behind the vegetation. Then at $t = 4.7$ sec and $t = 4.8$ sec (Figure 7b and c, respectively), the wave propagates in between the cylinders with maximum velocity and as a result, the wake behind the structures increases and minor reflections start to form. Once the wave has moved further downstream from the vegetation array, at $t = 4.9 - 5.1$ sec (Figure 7d-f) the fluctuations around the cylinders increase and large reflections are observed as a result of the oscillating motion of the wave.

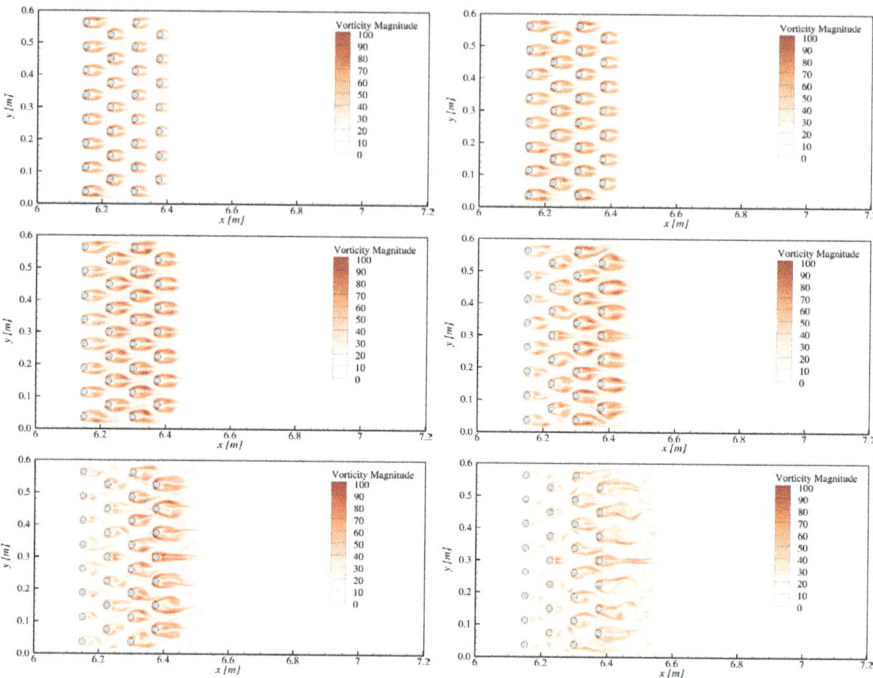

Fig. 8. Snapshots for the vorticity field during wave-cylinder-array interaction using the Hydro3D code.[54]

Figure 8 plots contours of the vorticity magnitude in $x - y$ plane at $z = 0.2$m visualising the flow and vortex shedding downstream of the vegetation. At $t = 4.6$ sec (Figure 8a) flow separation occurs and small eddies are formed as a result of the location of the wave in regard to the cylinders. Moreover, it is observed that the length and intensity of the vortex shedding as well as the boundary layer decrease at subsequent downstream locations due to higher velocities near the crest of the wave. At later time steps ($t = 4.7 - 4.8$ sec) the vorticity intensity and the overall length of the flow separation increase as the wave propagates through the vegetation array. In these instances, the flow around the cylinders at the centre of the flume is mostly symmetrical whereas a non-symmetrical flow is observed near the lateral boundaries due to side wall effects. At the remaining time steps ($t = 4.9 - 5.1$ sec) the flow is mostly random and the symmetrical vorticity no longer exists and a more complex flow field is formed due to the interaction of the solitary wave with the reflected waves and the cylinders.

3.4. *Wave-flexible structure interaction*

For wave-flexible-vegetation interaction, normally the fluid is described in an Eulerian formulation on a grid while the structures are presented through Lagrangian points or FEM formulations. In mesh-based method, flexible vegetation can be considered as a slender rod[36] or a beam[43] model. Instead of numerical methods based on Cartesian grids (as in the examples presented above), there is also the Lagrangian Smooth Particle Hydrodynamics (SPH) method, in which both the fluid and solid particles are solved in a Lagrangian way. In this approach, the same governing equations can be used for both fluid and solid particles, in which the time evolution of the deviatoric shear stress in solid follows the corrected Hooke's law. The deformation of the vegetation is considered by solid elasticity, which allows a fully coupled fluid-structure interaction simulation.[46] In the SPH method, there is no requirement for water-air interface calculation as all fluid particles (including the water surface) are tracked in the simulation as Lagrangian particles.

[46] first validated solid elasticity by considering a benchmark of free oscillations of a thin plate. After that, 2D simulations were performed for method calibration and compared with experimental results for flow vegetation interaction with plants mimicking Posidonia oceanica.[12] Then 3D simulation were carried out for vegetation with real Zostera noltei and then the numerical results were used to evaluate Manning's n for a great number of plants used for coastal management.

Figure 9 shows a recent example of the movement of flexible vegetation under waves at different water depth, in which waves are interacting with 200 emergent plants as one meadow. The top of Figure 9 shows a single meadow whereas the bottom of that figure shows two separate submerged meadows. In these 3D simulations, different particle sizes can be used for the water waves and the vegetation. However, due to the fact that the size of the actual plant is very small, these were enlarged by two orders of magnitude in the simulation to reduce the computational effort. It can be seen from Figure 9 that the phase and shape of leaf movements during WVI can be reproduced in the SPH simulation and it agrees qualitatively with laboratory and field observations. More research is needed to further our understanding of turbulence structures under wave-flexiable vegetation interaction.

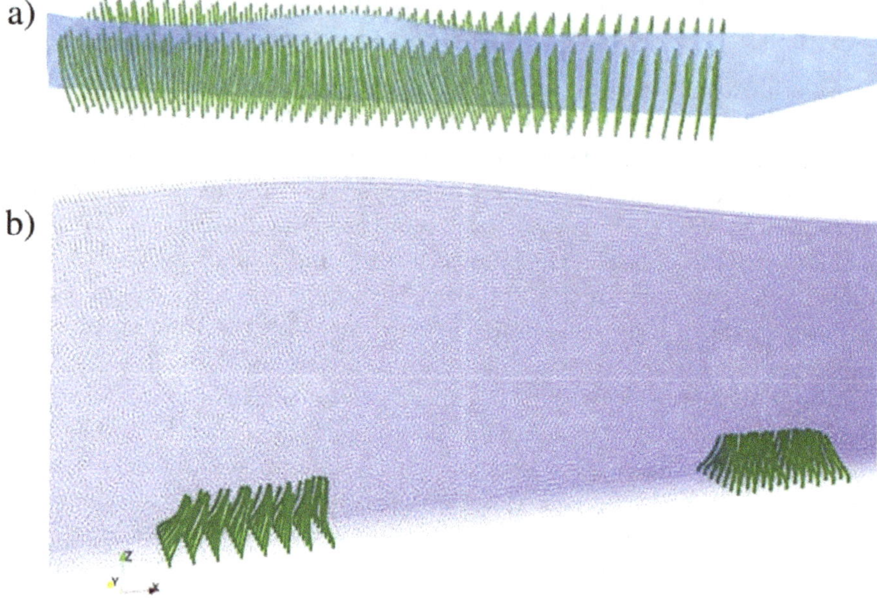

Fig. 9. Movement of the seagrass meadow under waves at different depths using the GPUSPH. (The picture is taken from[46]).

4. Closure

In this chapter numerical methods to facilitate wave-vegetation interaction have been introduced and discussed. Most studies to date on this or similar (e.g. uniform flow through vegetation or wave-cylinder interaction) have been based on the 2D or 3D Navier-Stokes equations together with turbulence closures including Reynolds-averaged (RANS) approaches or even the eddy-resolving method of LES. Clearly, as the complexity of the problem at hand increases, the demands of the numerical model to be able to reproduce accurately the physics of WVI by rises exponentially. WVI is governed by mechanisms taking place at a wide range of scales ranging from small-scale fluid separation at the plant stem scale (of the order of millimeters), the motion of the plant (if it is flexible) at the scale equivalent with the plant's maximum deflection to the scale of the water depth and if vegetation arrays are considered at the array scale which can be of the order of kilometers. At this moment in time computers are not powerful enough to allow simulations to cover all scales. At field scale (kilometers) WVI is simplified in computer models and the effect of vegetation on the fluid and the waves is accounted for by subgrid closure models (e.g. by employing the drag force concept). At smaller (mainly laboratory) scale, real vegetation is replaced with surrogates (to date mainly rigid cylinders or flexible plastic strips) and sophisticated RANS simulations and LES have revealed important physics of WVI. Some examples of such approaches were provided in this chapter. As computing power increases, such methods may become practical but this may take at least another 10 years.

References

1. C.-W. Chang and N. Mori, Green infrastructure for the reduction of coastal disasters: a review of the protective role of coastal forests against tsunami, storm surge, and wind waves, *Coastal Engineering Journal.* **63**(3), 370–385 (2021).
2. Y. Ozeren, D. G. Wren, and W. Wu, Experimental investigation of wave attenuation through model and live vegetation, *Journal of Waterway Port Coastal and Ocean Engineering.* **140**(5) (2014).
3. M. Maza, J. L. Lara, and I. J. Losada, Solitary wave attenuation by vegetation patches, *Advances in Water Resources.* **98**, 159–172 (2016).
4. F. He, J. Chen, and C. B. Jiang, Surface wave attenuation by vegetation with the stem, root and canopy, *Coastal Engineering.* **152** (2019).
5. J. R. Lei and H. Nepf, Wave damping by flexible vegetation: Connecting individual blade dynamics to the meadow scale, *Coastal Engineering.* **147**, 138–148 (2019).
6. N. H. Ram, V. Sriram, and K. Murali, Experimental investigation on the characteristics of solitary and elongated solitary waves passing over vegetation belt, *Journal of Ocean Engineering and Marine Energy.* **8**(3), 305–318 (2022).
7. M. Maza, J. L. Lara, I. J. Losada, B. Ondiviela, J. Trinogga, and T. J. Bouma, Large-scale 3-d experiments of wave and current interaction with real vegetation. part 2: Experimental analysis, *Coastal Engineering.* **106**, 73–86 (2015).
8. S. Lou, M. Chen, G. F. Ma, S. G. Liu, and G. H. Zhong, Laboratory study of the effect of vertically varying vegetation density on waves, currents and wave-current interactions, *Applied Ocean Research.* **79**, 74–87 (2018).
9. Z. Hu, S. Lian, T. Zitman, H. Wang, Z. He, H. Wei, L. Ren, W. Uijttewaal, and T. Suzuki, Wave breaking induced by opposing currents in submerged vegetation canopies, *Water Resources Research.* **58**(4) (2022).
10. T. J. van Veelen, T. P. Fairchild, D. E. Reeve, and H. Karunarathna, Experimental study on vegetation flexibility as control parameter for wave damping

and velocity structure, *Coastal Engineering.* **157** (2020).

11. X. Tang, P. Lin, Y.-H. Lin, Y. Jiang, and P. L.-F. Liu, Turbulence kinetic energy inside suspended vegetation domain under periodic water waves, *Journal of Waterway, Port, Coastal, and Ocean Engineering.* **149**(3), 04023004 (2023).

12. M. Luhar and H. M. Nepf, Flow-induced reconfiguration of buoyant and flexible aquatic vegetation, *Limnology and Oceanography.* **56**(6), 2003–2017 (2011).

13. K. C. Riffe, S. M. Henderson, and J. C. Mullarney, Wave dissipation by flexible vegetation, *Geophysical Research Letters.* **38** (2011).

14. H. M. Nepf, Flow and transport in regions with aquatic vegetation, *Annual Review of Fluid Mechanics.* **44**(1), 123–142 (2012).

15. R. Dalrymple, J. Kirby, and P. Hwang, Wave diffraction due to areas of energy dissipation, *Journal of Waterway Port Coastal and Ocean Engineering-Asce.* **110**, 67–79 (1984).

16. N. Kobayashi, A. W. Raichle, and T. Asano, Wave attenuation by vegetation, *Journal of Waterway Port Coastal and Ocean Engineering-Asce.* **119**(1), 30–48 (1993).

17. F. J. Mendez and I. J. Losada, An empirical model to estimate the propagation of random breaking and nonbreaking waves over vegetation fields, *Coastal Engineering.* **51**(2), 103–118 (2004).

18. I. J. Losada, M. Maza, and J. L. Lara, A new formulation for vegetation-induced damping under combined waves and currents, *Coastal Engineering.* **107**, 1–13 (2016).

19. C. C. Mei, I. C. Chan, P. L. F. Liu, Z. H. Huang, and W. B. Zhang, Long waves through emergent coastal vegetation, *Journal of Fluid Mechanics.* **687**, 461–491 (2011).

20. X. C. Tang, P. L. F. Liu, P. Z. Lin, Y. Jiang, and Y. H. Lin, An empirical model for predicting wave attenuation inside vegetation domain, *Ocean Engineering.* **257** (2022).

21. N. G. Jacobsen and B. C. McFall, Wave-averaged properties for non-breaking waves in a canopy: Viscous boundary layer and vertical shear stress distribution, *Coastal Engineering.* **174** (2022).

22. T. Suzuki, M. Zijlema, B. Burger, M. C. Meijer, and S. Narayan, Wave dissipation by vegetation with layer schematization in swan, *Coastal Engineering.* **59**(1), 64–71 (2012).

23. J. Tang, S. D. Shen, and H. Wang, Numerical model for coastal wave propagation through mild slope zone in the presence of rigid vegetation, *Coastal Engineering.* **97**, 53–59 (2015).

24. Y. Yao, Z. J. Tang, C. B. Jiang, W. R. He, and Z. S. Liu, Boussinesq modeling of solitary wave run-up reduction by emergent vegetation on a sloping beach, *Journal of Hydro-Environment Research.* **19**, 78–87 (2018).

25. Z. Y. Yang, J. Tang, and Y. M. Shen, Numerical study for vegetation effects on coastal wave propagation by using nonlinear boussinesq model, *Applied Ocean Research.* **70**, 32–40 (2018).

26. W. M. Wu and R. Marsooli, Adepth-averaged 2d shallow water model for breaking and non-breaking long waves affected by rigid vegetation, *Journal of Hydraulic Research.* **50**(6), 558–575 (2012).

27. J. Tang, Y. M. Shen, D. M. Causon, L. Qian, and C. G. Mingham, Numerical study of periodic long wave run-up on a rigid vegetation sloping beach, *Coastal Engineering.* **121**, 158–166 (2017).

28. T. Suzuki, Z. Hu, K. Kumada, L. K. Phan, and M. Zijlema, Non-hydrostatic modeling of drag, inertia and porous effects in wave propagation over dense vegetation fields, *Coastal Engineering.* **149**, 49–64 (2019).

29. A. A. van Rooijen, R. T. McCall, J. S. M. V. de Vries, A. R. van Dongeren, A. J. H. M. Reniers, and J. A. Roelvink, Modeling the effect of wave-vegetation interaction on wave setup, *Journal of Geophysical Research-Oceans.* **121**(6), 4341–4359 (2016).

30. G. F. Ma, J. T. Kirby, S. F. Su, J. Figlus, and F. Y. Shi, Numerical study of turbulence and wave damping induced by vegetation canopies, *Coastal Engineering.* **80**, 68–78 (2013).

31. J. X. Zhang, J. Wang, X. Fan, and D. F. Liang, Numerical investigation of water wave near-trapping by rigid emergent vegetation, *Journal of Hydro-Environment Research.* **25**, 35–47 (2019).

32. K. Qu, G. Y. Lan, W. Y. Sun, C. B. Jiang, Y. Yao, B. H. Wen, Y. Y. Xu, and T. W. Liu, Numerical study on wave attenuation of extreme waves by emergent rigid vegetation patch, *Ocean Engineering.* **239** (2021).

33. R. Familkhalili and N. Tahvildari, Computational modeling of coupled waves and vegetation stem dynamics in highly flexible submerged meadows, *Advances in Water Resources.* **165** (2022).

34. C. W. Li and M. L. Zhang, 3d modelling of hydrodynamics and mixing in a vegetation field under waves, *Computers & Fluids.* **39**(4), 604–614 (2010).

35. C. W. Li and K. Yan, Numerical investigation of wave-current-vegetation interaction, *Journal of Hydraulic Engineering.* **133**(7), 794–803 (2007).

36. H. Chen and Q.-P. Zou, Eulerian–lagrangian flow-vegetation interaction model using immersed boundary method and openfoam, *Advances in Water Resources.* **126**, 176–192 (2019).

37. R. Marsooli and W. M. Wu, Numerical investigation of wave attenuation by vegetation using a 3d rans model, *Advances in Water Resources.* **74**, 245–257 (2014).

38. A. Chakrabarti, Q. Chen, H. D. Smith, and D. Liu, Large eddy simulation of unidirectional and wave flows through vegetation, *Journal of Engineering Mechanics.* **142**(8) (2016).

39. Q. Chen and H. H. Zhao, Theoretical models for wave energy dissipation

caused by vegetation, *Journal of Engineering Mechanics.* **138**(2), 221–229 (2012).

40. S. Hadadpour, M. Paul, and H. Oumeraci, Numerical investigation of wave attenuation by rigid vegetation based on a porous media approach, *Journal of Coastal Research.* pp. 92–100 (2019).

41. T. J. van Veelen, H. Karunarathna, and D. E. Reeve, Modelling wave attenuation by quasi-flexible coastal vegetation, *Coastal Engineering.* **164** (2021).

42. M. Maza, J. L. Lara, and I. J. Losada, A coupled model of submerged vegetation under oscillatory flow using navier–stokes equations, *Coastal Engineering.* **80**, 16–34 (2013).

43. S. A. Mattis, C. E. Kees, M. V. Wei, A. Dimakopoulos, and C. N. Dawson, Computational model for wave attenuation by flexible vegetation, *Journal of Waterway Port Coastal and Ocean Engineering.* **145**(1) (2019).

44. R. B. Zeller, F. J. Zarama, J. S. Weitzman, and J. R. Koseff, A simple and practical model for combined wave-current canopy flows, *Journal of Fluid Mechanics.* **767**, 842–880 (2015).

45. R. Divya, V. Sriram, and K. Murali, Wave-vegetation interaction using improved meshless local petrov galerkin method, *Applied Ocean Research.* **101** (2020).

46. A.-E. Paquier, T. Oudart, C. Le Bouteiller, S. Meulé, P. Larroudé, and R. A. Dalrymple, 3d numerical simulation of seagrass movement under waves and currents with gpusph, *International Journal of Sediment Research.* **36**(6), 711–722 (2021).

47. T. Stoesser, G. P. Salvador, W. Rodi, and P. Diplas, Large eddy simulation of turbulent flow through submerged vegetation, *Transport in Porous Media.* **78**(3), 347–365 (2009).

48. T. Stoesser, S. J. Kim, and P. Diplas, Turbulent flow through idealized emergent vegetation, *Journal of Hydraulic Engineering.* **136**(12), 1003–1017 (2010).

49. N. G. Jacobsen, D. R. Fuhrman, and J. Fredsøe, A wave generation toolbox for the open-source cfd library: Openfoam, *International Journal for Numerical Methods in Fluids.* **70**(9), 1073–1088 (2012).

50. P. Higuera, J. L. Lara, and I. J. Losada, Realistic wave generation and active wave absorption for navier–stokes models: Application to openfoam®, *Coastal Engineering.* **71**, 102–118 (2013).

51. P. Lin and P. L.-F. Liu, Internal wave-maker for Navier-Stokes equations models, *Journal of waterway, port, coastal, and ocean engineering.* **125**(4), 207–215 (1999).

52. P. Higuera, I. J. Losada, and J. L. Lara, Three-dimensional numerical wave generation with moving boundaries, *Coastal Engineering.* **101**, 35–47 (2015).

53. Z. Xie, S. Lopez, A. Christou, T. Stoesser, L. Lu, and P. Lin, Numerical investigation of steep focused wave interaction with a vertical cylinder using

a Cartesian cut-cell method, *International Journal of Offshore and Polar Engineering.* **31**, 78–86 (2021).

54. A. Christou, T. Stoesser, and Z. Xie, A large-eddy-simulation-based numerical wave tank for three-dimensional wave-structure interaction, *Computers & Fluids.* **231**, 105179 (2021).

55. Z. Xie and P. Lin, Eulerian and lagrangian transport by shallow-water breaking waves, *Physics of Fluids.* **34**(3), 032116 (2022).

56. Z. H. Huang, Y. Yao, S. Y. Sim, and Y. Yao, Interaction of solitary waves with emergent, rigid vegetation, *Ocean Engineering.* **38**(10), 1080–1088 (2011).

57. L. Zhu and Q. Chen, Numerical modeling of surface waves over submerged flexible vegetation, *Journal of Engineering Mechanics.* **141**(8) (2015).

58. K. Yin, S. D. Xu, W. R. Huang, S. Liu, and M. X. Li, Numerical investigation of wave attenuation by coupled flexible vegetation dynamic model and xbeach wave model, *Ocean Engineering.* **235** (2021).

59. M. Uhlmann, An immersed boundary method with direct forcing for the simulation of particulate flows, *Journal of Computational Physics.* **209**(2), 448 476 (2005).

60. Y.-H. Tseng and J. H. Ferziger, A ghost-cell immersed boundary method for flow in complex geometry, *Journal of Computational Physics.* **192**(2), 593–623 (2003).

61. Z. Xie, An implicit cartesian cut-cell method for incompressible viscous flows with complex geometries, *Computer Methods in Applied Mechanics and Engineering.* **399**, 115449 (2022).

62. J. Yang and F. Stern, Sharp interface immersed-boundary/level-set method for wave-body interactions, *Journal of Computational Physics.* **228**(17), 6590–6616 (2009).

63. Z. Xie and T. Stoesser, A three-dimensional Cartesian cut-cell/volume-of-fluid method for two-phase flows with moving bodies, *Journal of Computational Physics.* **416**, 109536 (2020).

64. F. Sotiropoulos and X. Yang, Immersed boundary methods for simulating fluid–structure interaction, *Progress in Aerospace Sciences.* **65**, 1–21 (2014).

65. C. Y. H. Wong, A. S. Dimakopoulos, P. H. Trinh, and S. J. Chapman, Multiple-scales analysis of wave evolution in the presence of rigid vegetation, *Journal of Fluid Mechanics.* **935** (2022).

66. W. Rodi, G. Constantinescu, and T. Stoesser, *Large-Eddy Simulation in Hydraulics.* CRC Press, Boca Raton, Florida, U. S. (2013).

67. T. Stoesser, Large-eddy simulation in hydraulics: Quo vadis?, *Journal of Hydraulic Research.* **52**, 441–452 (2014).

68. W. H. Mo, K. Irschik, H. Oumeraci, and P. L. F. Liu, A 3D numerical model for computing non-breaking wave forces on slender piles, *Journal of Engineering Mathematics.* **58**(1-4), 19–30 (2007).

69. Z. Xie, T. Stoesser, S. Yan, Q. Ma, and P. Lin, A Cartesian cut-cell based

multiphase flow model for large-eddy simulation of three-dimensional wave-structure interaction, *Computers & Fluids.* **213**, 104747 (2020).

70. S. J. Kim and T. Stoesser, Closure modeling and direct simulation of vegetation drag in flow through emergent vegetation, *Water Resources Research.* **47**(10) (2011).

71. A. Christou, D. Stagonas, E. Buldakov, and T. Stoesser, Focused waves on shear currents interacting with a vertical cylinder, *Coastal Engineering.* **231**, 105179 (2023).

72. J.-J. Lee, J. E. Skjelbreia, and F. Raichlen, Measurement of velocities in solitary waves, *Journal of the Waterway, Port, Coastal and Ocean Division.* **108**(2), 200–218 (1982).

Chapter 9

A Comprehensive Review on Wave–Vegetation Interaction

N. Hari Ram[1] and V. Sundar[2]

[1]*Research Scholar at Department of Ocean Engineering, Indian Institute of Technology Madras, Chennai, India – 600 036*

[2]*Advisory Consultant, Department of Ocean Engineering, Indian Institute of Technology Madras, Chennai, India – 600 036*
vallamsundar@gmail.com

The coastal vegetation is now closely regarded as one of the nature-based solution for protecting the coast against hazards such as storm surges and tsunamis. This chapter presents a comprehensive review on the interaction of vegetation with extreme flow due to the hazard as mentioned above through physical, numerical and field studies. Signature studies from the post Great Indian Ocean tsunami revealed the effectiveness of vegetation as a measure of reducing the inundation distance and height. This opened a new dimension forcing researchers to examine the wave-vegetation interaction problem. Conventional hard measures such as seawalls, bulkheads and revetments have been used successfully as coastal protection, although negative effects due to such structures have also been reported in the literature. The vegetation is considered as an eco-friendly measure that plays both the role of coastal protection and in ensuring the safety of the habitat if planned and managed effectively.

1. Tsunami

Tsunamis are caused due to underwater volcanic eruption, earthquakes or landslides which transform into bores when they reach the coast and are one of the most dangerous threats. They are long-period waves that can travel hundreds of kilometers and poses the characteristics of shallow water waves even in the very deep waters. Tsunamis inundate the coast with massive destructive energy leading to loss of humanity and property [Rodriguez *et al.*, [2006]; Sheth *et al.*, [2006] Mimura *et al.*, [2011]; Palermo *et al.*, [2013]; Mikami *et al.*, [2019]].

Several tsunamis have occurred in the past, the details of all such events are not available. Soon after the great Indian Ocean tsunami of 2004, Gusiakov [2009] reported the details of the 1965 different tsunamis from 2000 BC along with their respective sources, as can be seen in Figure 1. Tinti *et al.*, [2006] has reported the details of the two tsunamis that triggered due to landslides in Stromboli, southern Tyrrhenian Sea, Italy in December 2002. The two tsunamis occurred within an interval of 7 mins. Comparing the run-up between the two tsunamis the former was observed to be relatively higher.

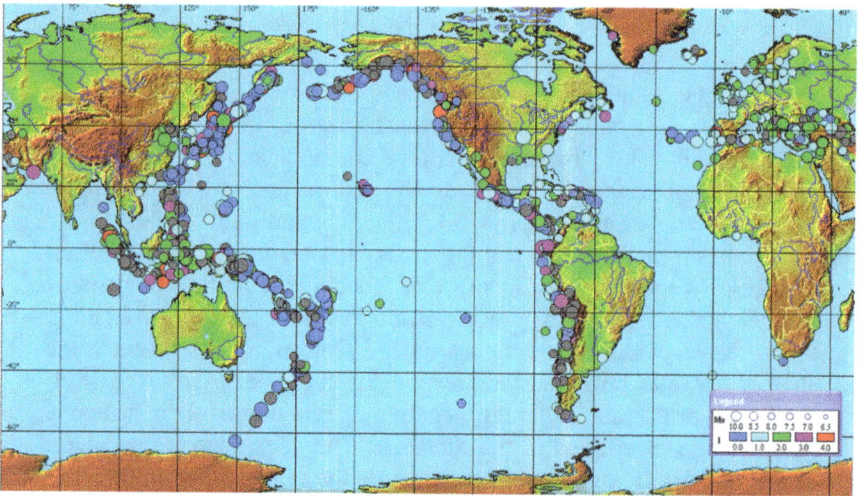

Fig. 1. Visualization of the NTL/ICMMG global historical tsunami catalog. 1965 tsunamigenic events with identified sources are shown for the period from 2000 BC to present time. Size of circles is proportional to event magnitude (for seismically induced tsunamis), color represents tsunami intensity on the Soloviev – Imamura scale [Gusiakov, 2009].

The subduction of the Indian plate under the Burma microplate resulted in a massive earthquake of magnitude 9.3 that served as a trigger for the 2004 Great Indian Ocean tsunami that propagated across the globe affecting the coasts of India including Andaman and Nicobar Islands,

Sri Lanka, Sumatra, Malaysia, and Thailand. Sundar *et al.*, [2007] reported in detail the variation of run-up along various parts of the Indian coast and reasons behind the failure of structures. Lavigne *et al.*, [2007] detailed the 2006 Java island tsunami caused by an earthquake of 7.8 magnitude that lead to the generation of three consecutive waves with a maximum flow depth of about 5 m at the coasts which had resulted in the loss 730 lives. The reports from Fritz [2007] indicated a run-up range of 5 to 7 m along the Java coast, whereas, at a few locations along the south coast of Nusa Kambangan, it was reported to be up to 21 m. The 2010 tsunami in Chile generated due to an earthquake of magnitude 8.8 caused a death toll of more than 170 human lives.

The 2009 Samoan Island tsunami was also generated due to an earthquake that occurred in the south central Pacific south to Samoan island. The preliminary survey investigation done by Okal *et al.*, [2010] reported that the maximum run up was about 17 m. During the same year, coasts of Sumatra also faced a tsunami impact which was caused by an earthquake of magnitude 7.7 that resulted in more than 500 causalities. The tsunami heights were found mostly to be in the range of 4 to 7 m with an inundation more than 300 m as reported by Satake *et al.*, [2013]. Following the 2004 Great Indian Ocean tsunami the 2011 Tokyo tsunami had the catastrophic impact. Earthquake of magnitude 9.0 resulted in the generation of tsunami that had run up around the range of 5–10 m as reported by Fritz *et al.*, [2011]. An Earthquake of magnitude 7.5 resulted in a tsunami that struck the coast of Indonesia on September 2018. Mikami *et al.*, [2019] reported that the inundations were observed to be up-to 6.77 m. Dogan *et al.*, [2021] reported a maximum run up of 3.8 m at a distance 91 m from the shore line by a tsunami that was generated by an earthquake of magnitude 6.6 in the Aegean Sea tsunami 2020. Han and Yu [2022] reported the details of the 2022 Tonga tsunami due to a volcanic eruption south west Pacific. The amplitude of the tsunami ranged from 0.8 to 1.7 m in the coastal areas of Tonga. Table 1 has list of the tsunamis of the 20th century and the cause that triggered their generation. It can be seen from the table and literature that the most of the destructive tsunami is caused by earthquakes.

Irish *et al.*, [2012] has reported the significant role of topography on the 2009 Samoa tsunami. The landscape in a few villages was severely

Table 1. List of 20th century tsunami and its cause of trigger

Year	Cause	Location of impacts
2002	Landslide	Tyrrhenian sea
2004	Earthquake	Indonesia, Sri Lanka, India, Maldives & Thailand
2006	Earthquake	Java Islands
2006	Earthquake	Kuril Islands
2007	Earthquake	Solomon Islands
2007	Earthquake & landslide	Chile
2007	Landslide	Columbia
2009	Earthquake	Samoa
2010	Earthquake	Chile
2010	Earthquake	Sumatra
2011	Earthquake	Japan
2012	Earthquake	El Salvador & Nicaragua
2013	Earthquake	Solomon Islands
2014	Landslide	Iceland
2015	Earthquake	Chile
2015	Earthquake	Taan Fiord, Alaska, & U.S.
2016	Earthquake	New Zealand
2017	Landslide	Greenland
2018	Landslide	Sulawesi
2018	Volcanic eruption & landslide	Java and Sumatra
2020	Earthquake	Aegean Sea
2020	Landslide	Canada
2021	Earthquake	Norfolk Islands & New Zealand
2021	Earthquake & landslide	Ambon
2021	Earthquake	Alaska
2021	Earthquake	South Atlantic
2022	Volcanic eruption	Tonga
2022	Earthquake	Mexico
2022	Landslide	Philippines

destroyed due to the confined nature of headlands and the slope surrounding them. Muhari *et al.*, [2018] investigated and reported the details of the 2018 Indonesia tsunami and its impact. The inundation depth was recorded to be 8 m with an inundation distance of about 300 m. Thus, protecting the coasts with sustainable and eco-friendly measures needs to be planned along with the coastal infrastructure development. The following sections provide an insight into the factors to be considered while, planning to setting up vegetation as a part of coastal protection measures against tsunamis and other extreme coastal hazards.

2. Mitigation measures

Coastal protection against perineal erosion and to combat extreme events like storm surges and tsunamis is vital. The impact 2004 tsunami is one of the deadliest tsunamis, [Rodriguez *et al.*, 2006] the impact of which on human lives and to the structures was most unexpected. Later from the field investigation, the role of coastal vegetation in protecting the hinterland had gained special attention. The mangrove trees and coastal reefs behaved as natural wave attenuators. The coastal vegetation is being proposed to be an eco-friendly protection measure by several researchers. Figure 2a shows the areas of coastal vegetation across the globe. It is reported that the area extent of mangrove forests in South Asia is approximately 1,187,476 ha representing 7% of the global total. India has 4500 square kilometers of mangroves. The majority (60%) is found in the east coast compared to the west coast (14%) due to the nutrient rich deltas and suitable terrain. The remaining 26% are found in the Andaman and Nicobar Islands. West Bengal has the largest area of mangroves, followed by Gujarat, Andaman and Nicobar. The effects of the tsunami of 2004 along the coasts of India have been reported by Sheth *et al.*, [2006] and Sundar *et al.*, [2007]. The devastating impact of the 2011 Tohoku tsunami in Japan was reported by Mimura *et al.*, [2011]. Palermo *et al.*, [2013] reported the impacts due to 2010 Chile tsunami. The facts after field investigation aftermath of tsunami reported by [Danielsen *et al.*, 2005; Sundar *et al.*, 2007] revealed that the existence of coastal vegetation has relatively reduced the impact of tsunami. The tsunami prone coastal regions around the globe including the "Ring of Fire" with active underwater volcanoes are shown in Figure 2b. The effect of vegetation in reducing the impact of tsunami 2004 along the south east coast of Tamil Nadu is shown in Figure 3.

If one compares the coastal vegetation distribution map and the tsunami risk map shown in the above figures, proving the importance of

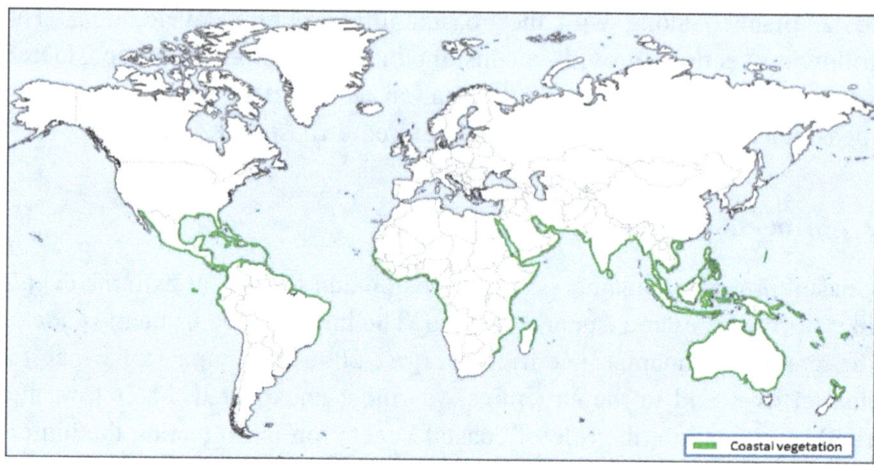

Fig. 2a. Coastal vegetation covers across world.

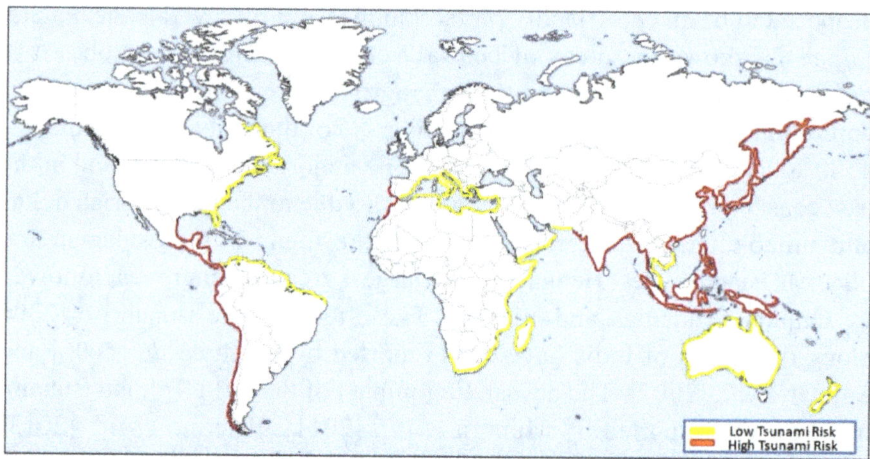

Fig. 2b. Tsunami prone coastal areas across world.

natural coastal vegetation. Eco-system based coastal protection is a concept that enhances the ability of the natural environment to protect the coastal community and the infrastructure present in that coastal region. The impact of coastal hazards associated with climate change on human livelihood and infrastructure has become a challenge to plan for countermeasures. Duarte *et al.*, [2013] has mentioned the key role of coastal vegetation in reducing the effects of the current scenario of the vulnerability of coast due to climate change. Coastal vegetation also helps in protecting against soil erosion which has been scientifically investigated by Thampanya *et al.*, [2006]; Montakhab *et al.*, [2012]; Besset *et al.*, [2019] and shown its effectiveness. It also helps in carbon sequestration as reported by Breithaupt *et al.*, [2012], Bianchi *et al.*, [2013], Ward [2020]. A detailed survey reveals the effect of coastal vegetation in protecting the hinterlands as discussed by Tanaka, [2007]; Alongi, [2008]; Mascarenhas and Jayakumar, [2008]; Tanaka, [2009]; Feagin *et al.*, [2010]; Cochard, [2011]; Gedan *et al.*, [2011]; Marois and Mitsch, [2015]; and Chang and Mori, [2021]. Sundar *et al.*, [2020] have detailed the Engineering Perspective for Mitigation, Protection and Modelling of tsunamis. The present paper complements the discussion and provides an insight to the aspects of coastal protection with vegetation and aims to give a comprehensive knowledge on the wave-vegetation interaction.

Fig. 3. Role of vegetation during the tsunami 2004.

3. Field observations

The field investigations bring in the actual post scenario of any event. The recent technological advancement aids in understanding the impacts and planning for future events. The field survey by the researchers revealed the importance of vegetation acting as a natural buffer protecting the areas behind it from tsunamis. Shuto, [1987] through an analysis considering five major tsunamis that struck Japan has discussed the effectiveness of vegetation in reducing the adverse effects of tsunami impact. Through the study, monograms were proposed, which gives a general estimate of the tree breast diameter required to be effective over a range of inundation. Chang *et al.*, [2006] through remote sensing analysis of the 2004 tsunami event in the North Andaman coasts of Thailand has stated that the communities established behind mangroves sustained minimal damages. The assessments were reported based on Papadopoulos and Imamura tsunami intensity scale. Figure 4 shows the damage that occurred due to the tsunami to the buildings marked by circles. These were considered by Suppasri *et al.*, [2011] from Jan 2003 and Jan 2005 satellite images. It can be observed that due to urbanization, vegetation being removed causing the buildings exposed directly to the coastal hazards being severely damaged compared to other buildings sheltered by vegetation.

Fig. 4. Satellite image of coastal building before and after 2004 Tsunami [Suppasri *et al.*, 2011].

Apart from tsunamis, there are reports that the coastal forest protects coasts from storm surges. Das and Vincent, [2009] has reported the effectiveness of the mangroves in reducing the death toll during a super cyclone struck in 1999. Krauss *et al.*, [2009] study on 2004 and 2005 hurricane events along the coast of Florida observed the protective role of the mangroves. It has been reported that there was a reduction of 4.2 and 9.4 cm/km in a riverine mangrove and interior mangrove marshes, respectively. Other researchers later tested these aspects of vegetation in its role in attenuating the ocean waves through physical and numerical modelling, as mentioned in the following sections in detail.

4. Experimental investigations

Well controlled experimental investigations are crucial to understand the physics behind the behavior of vegetation under extreme coastal hazards and to quantity its effectiveness in attenuating the flow velocity and its distance of inundation. It also helps in quantifying the parameters that are involved in the wave-vegetation interaction, a topic, on which several studies have been done in the past. The parameters include density, submergence ratio, incident wave height, vegetation arrangement, the diameter of the stem of the vegetation, species, etc., which influence the wave attenuation and energy reduction has been proven. Figure 5 provides an overall picture of the physical aspects and the importance of vegetation protecting the coast.

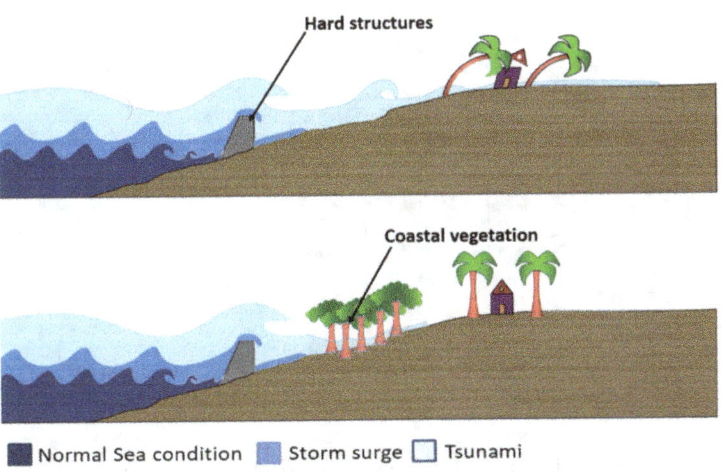

Fig. 5. Protection provided by vegetation from extreme events.

The parameters involved in hydrodynamic studies can be measured and clearly understood. However, on the wave-vegetation interaction, the parameters mentioned under the vegetation are interdependent. They all contribute a part to each of the parameters involved with the vegetal characteristics. For instance, the drag force and vortex shedding effects are influenced by the diameter of the trunk, root system, arrangement of vegetation and density. The resistance offered by the vegetation belt is dependent on the width of the vegetation belt, density, submergence of the vegetation and stiffness. All these parameters also vary depending on the hydrodynamic flow interacting with the vegetation. An overall picture of the parameters involved in the wave vegetation interaction study is illustrated in Figure 6. Thus, various investigations have been taken up by researchers which are in progress.

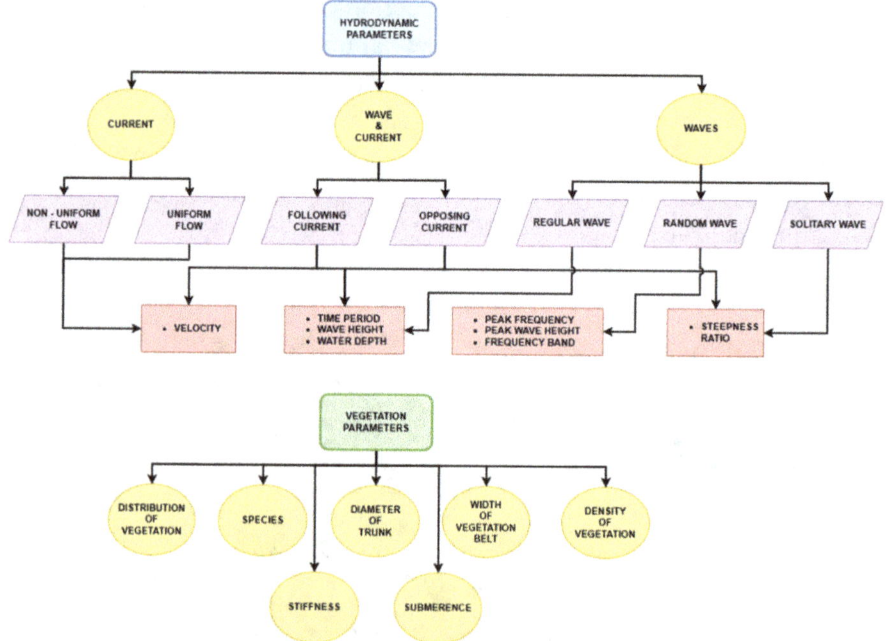

Fig. 6. Parameters involved in wave–vegetation interaction study.

The protection offered by the vegetation is mainly through drag force. Dalrymple *et al.*, [1984] proposed an analytical model for the drag force for vegetation field on a horizontal bed. The intensity of drag induced by roots, stem region, and the canopy region was different. Fischenich, [2000] has shown the variation of flow velocities for different flow conditions of submergence. Struve *et al.*, [2003] observed that the drag force increases with an increase in the modelled tree surface (i.e. diameter and density). Mendez and Losada, [2004] proposed empirical models for estimating drag force as an extension to Dalrymple *et al.*, [1984] considering sloping bed and breaking and non-breaking random waves over vegetation fields. The energy dissipation by vegetation due to vegetal parameters like density, stem diameter, and vegetation height has also been reported. The energy dissipation due to wave breaking over vegetation fields was also discussed. Noarayanan *et al.*, [2012a] studied the variation of the resistance offered by the vegetation for different spacing. Through the experimental data, an empirical equation for determining the flow resistance for the tandem arrangement of vegetation has been proposed. Noarayanan *et al.*, [2012b] through experimental study has proposed empirical equations for staggered arrangement of the stems. A Vegetation Flow Parameter *(VFP)* was proposed as a modified relative rigidity that included the vegetation parameters like spacing between the vegetation, diameter, width of the vegetation belt, the velocity of flow, flow-depth, the height of vegetation and rigidity of the vegetation. The vegetal resistance and drag were observed to correlate well with this parameter.

The drag force varies not only with different species but also with the depth of vegetation. Zhao and Fan, [2017] reported the flow resistance offered by submerged vegetation in terms of permeability, in which, truncated cones were considered as a model representing vegetation. Instead of the density of vegetation, the porosity was used to account for the shape of vegetation. It was observed that the resistance to flow was inversely proportional to the porosity. Wu and Cox, [2016] studied the wave attenuation characteristics of vegetation considering vertical vegetation. The damping and the drag coefficient of vegetation with varied vertical density was observed to be insignificant. However, the influence of this vertical density variation in wave attenuation was not significant when exposed to shallow water waves. However, in relatively deep-water waves, the wave attenuation was less for a non-varied density vegetation patch. Thus, the vertical density variation is significant to the flow characteristics of shallow water and deep-water waves. In real-time scenario, the vegetation being at the coast may not be subjected to deep-water wave condition.

The experimental observations of Lou *et al.*, [2018] also reported the similar phenomenon of wave attenuation for deep-water and shallow water waves for vegetation fields having varied vertical densities. The reason is explained as the intensity of the turbulence decreases in the regions of lower vegetation height when the water depth increases. The velocity profile variation also differs based on the submergence condition. The variation of the velocity in the presence of rigid and flexible vegetation under different submergence flow obstructing an uniform flow is depicted in Figure 7.

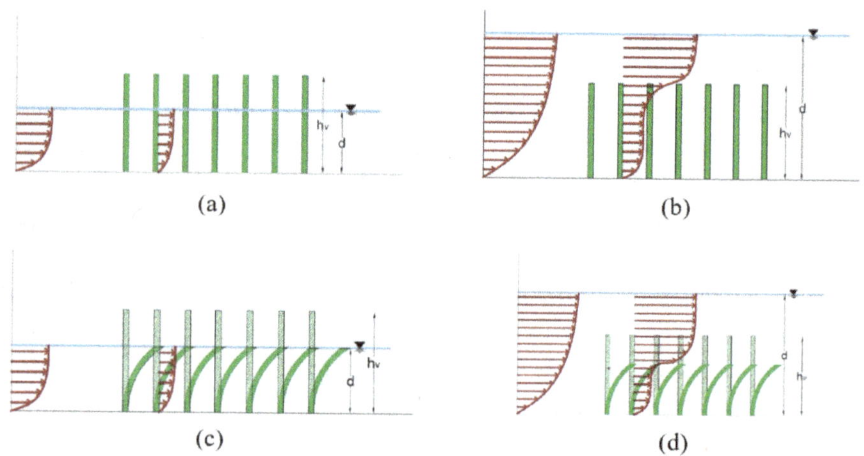

Fig. 7. Velocity variation for uniform flow in presence of a patch of rigid (a), (b) and flexible (c), (d) vegetation.

Panigrahi and Khatua, [2015] reported the effect of rigidity (flexible and stiff vegetation) on the drag induced by the vegetation. The rigid vegetation was observed to produce a higher drag compared to flexible vegetation. Lei and Nepf, [2019] studies with a flexible vegetation model observed that the arrangement of the vegetation has no significant effects in the wave decay. The drag produced by relatively stiff vegetation was higher for a given length of the leaf. This is due to the fact that the flexible vegetation moving passively with the wave. However, studies of Sundar *et al.*, [2011] with rigid vegetation showed that staggered and tandem arrangement of the stems was observed to be effective in reducing the pressures and run-up on a vertical wall. Wang *et al.*, [2022a] studied the wave damping characteristics considering artificial *Spartina alterniflora* model. The density, submergence ratio, relative wave height and wavelength were varied, and its influence on wave damping was reported. The observations stated that the density and the relative wavelength contributed more to the wave attenuation characteristics for the emergent condition compared to the submerged condition. Apart from the energy dissipation due to drag, the energy dissipation occurs due to turbulence-induced waves. Righetti, [2008] reported the presence of a higher transverse component of turbulent intensities within the vegetated field due to the Karman vortex shedding. This effect was also observed to be higher for dense vegetation. Zhang *et al.*, [2018] examined the variation in the turbulence kinetic energy *(TKE)* depending on the ratio of wave-excursion to stem spacing, in which, the *TKE* induced by rigid vegetation was observed to be high compared to that due to flexible vegetation.

Ishikawa *et al.*, [2000] reported that the drag coefficient from stems increases logarithmically with its density and its staggering pattern. Through the experimental investigation, relationships for the stem drag coefficient which appear useful in partitioning the total flow resistance of vegetated bed streams into the stem and the bed particle resistances were proposed. Busari *et al.*, [2013] studied the effects of velocity profile variation due to vegetation density for submerged vegetation. An increase in attenuation characteristics was observed due to an increase in the vegetation density. With an increase in vegetation density, the streamlining of the vegetation to the flow was also observed to be reduced, leading to a higher drag contribution from the vegetation. Strong shear layer formation has been reported at the tip of the vegetation.

Wu *et al.*, [1999] also reported a similar phenomenon. It was explained in terms of the boundary layer zone at the tip of the vegetation. In this study, the vegetation was represented with a rubberized horsehair mattress. Hiraishi [2003] investigated the effect of the density of vegetation in reducing the tsunami pressure and wave height. It was reported that 30 trees/m^2 has efficiently for tsunami waves of less than 4 to 5 m. The reduction of flow pressure of the vegetation belt was observed to be close to a coastal dike composed of rubble stones that was considered.

The vegetation morphology has a different root system, and its canopy distributions are unique to its species. Experimental research with real vegetation models has also been carried out by several researchers. Freeman *et al.*, [2000] investigated resistance and drag induced due to vegetation with 20 different plant species. Meijer and Van Velzer, [1999] investigated the applicability of the analytical roughness coefficient and the resulting flow profile with modelled rigid steel bars and natural Reed. The model, however, did not consider the effects of varying diameters, tufts, and the stiffens of the reed. The observations like canopy streamlining influenced by the velocity and the resistance influenced by submergence ratio were similar to the modelled vegetation of Fathi-Moghadam *et al.*, [2011]. The effects of vegetation canopy, roots and stem were reported by He *et al.*, [2019] by modelling *Rhizophora*. In general, the percentage of attenuation provided by the canopy was higher than the stem and root. However, in the case of limited water depth, the effect of roots was observed to be high. The investigations of Wang *et al.*, [2022b] on complex root morphology in mangrove forest revealed that the wave dissipation characteristics is influenced by the submergence of the roots. An empirical formula for estimating a bulk drag considering a modified KC number was proposed.

Irtem *et al.*, [2009] investigated the variation of run-up due to vegetation considering two different vegetation models and different arrangements and proposed empirical equations in terms of non-dimensional parameters. The relative position of the vegetation from the still water depth was found to influence the run-up. Lakshmanan *et al.*, [2012] reported that the position of a building of width greater than the width of the vegetation belt was observed to experience lesser force, which could be due to an increase in the reflection from the building with the incoming long wave not having enough region to dissipate.

Several studies investigated the effectiveness of vegetation in attenuating tsunamis being represented a solitary wave. The influence of the configuration of the vegetation and its width affecting the drag coefficient was reported by Huang *et al.*, [2011]. Maza *et al.*, [2016] considered different configurations of vegetation patches in wave basins and observed their influence on attenuation characteristics of solitary waves. Fathi-Moghadam *et al.*, [2018] investigated the effects of vegetation density, breaking velocity and the coastal slope in wave attenuation of a solitary wave. It was observed that the steeper the slope, the energy attenuation was high due to wave breaking. Tognin *et al.*, [2019] reported the influence of vegetation density on the velocity profile through flow visualization using Particle Image Velocimetry (PIV) for solitary waves passing over a rigid vegetation belt. Zhao *et al.*, [2021] study accounted for the effects of currents along with the solitary wave. It was reported that the solitary wave along with the positive current was readily attenuated by vegetation compared to the opposing current. Hari Ram *et al.*, [2022] carried out a comparative study on the vegetation interaction with solitary and elongated solitary waves. It was reported that the percentage of energy reduction remains the same for both models for a given vegetation width. A modified submergence ratio was proposed considering the flow characteristics of the solitary type waves over the vegetation belt. Ismail *et al.*, [2012] investigated run-up reduction due to vegetation considering the width of the forest and its density. In this study, the hydrodynamic characteristics of a tsunami were considered from dam-break flow characteristics. Considering two downstream water levels of dam break representing tidal water level observed the variation of run-up reduction. The interdependence of water level and the length of vegetation and the variation of run-up was observed to be similar to the findings of Irtem *et al.*, [2009]. Noarayanan *et al.*, [2012c] investigated the effect of vegetation belt in reducing run-up from regular and Cnoidal waves and reported that the reduction was influenced more by the width of the vegetation belt than its density.

Augustin *et al.*, [2009] studies exhibited that the emergent vegetation was effective in wave attenuation compared to that near emergent/ submerged vegetation, for regular waves and rigid vegetation models. In contrast, Hari Ram *et al.*, [2022] investigation with rigid vegetation showed the near emergent condition being more effective in wave atten-uation, while, considering solitary and elongated solitary waves. Similarly, Anderson and Smith, [2014] reported higher wave attenuation in the near emergent condition for long waves, in which, irregular waves interacting with flexible vegetation were also considered. It was also observed that the maximum energy loss occurs at the spectral peaks. Maza *et al.*, [2015a] considering real vegetation interacting with a regular wave with current

                                    N. Hari Ram and V. Sundar

reported that the wave attenuation is higher for a wave with an opposing current than the following current. Similar observation was made by Hu *et al.*, [2022] considering rigid vegetation model. The opposing current indirectly helped in the flow attenuation by inducing wave breaking inside the vegetation field in submerged conditions, whereas Zhao *et al.*, [2021] reported higher wave attenuation for solitary waves in following current compared to the opposing current.

The empirical equations related to the important flow parameters on experimental studies on wave vegetation interaction from a comprehensive literature review are provided in Table 2.

Table 2. Empirical equations based on different flow and the parameters considered by different investigators on wave vegetation studies

Authors	Flow condition	Formula
Dalrymple et al., (1984)	Regular wave Waves (Wave attenuation)	$\dfrac{a}{a_0} = \dfrac{1}{1+\alpha x}$
Wu et al., (1999)	Current (Drag coefficient)	$C_d = \dfrac{(3.44\times10^6)S^{0.5}}{R}$ for un-submerged vegetation $C_d = \dfrac{(4.50\times10^{10})S^{0.89}T^{1.09}}{R^{1.7}}$ for submerged vegetation
Freeman et al., (2000)	Current (Manning's coefficient)	$n = K_n 0.183 \left(\dfrac{E_s A_s}{\rho A_i V_*^2}\right)^{0.183} \left(\dfrac{H}{Y_o}\right)^{0.243} (MA_i)^{0.273} \left(\dfrac{v}{V_* R_h}\right)^{0.115} \left(\dfrac{1}{V^*}\right) R_h^{2/3} S^{1/2}$ for submerged vegetation $n = K_n 3.487E-05 \left(\dfrac{E_s A_s}{\rho A_i V_*^2}\right)^{0.15} (MA^*_i)^{0.273} \left(\dfrac{V_* R_h}{v}\right)^{0.622} \left(\dfrac{R_h^{2/3} S^{1/2}}{V^*}\right)$ for partially submerged vegetation
Mendez, & Losada (2004)	Irregular waves (Drag coefficient)	$C_d = 0.47exp(-0.052K)$ for $3 \le K \le 59$ $C_d = exp(-0.0138Q)Q^{0.3}$ for $7 \le Q \le 172$
Irtem et al., (2009)	Long waves (Run up)	$\dfrac{R}{d} = C1 \left(\dfrac{HG_{sb}(cot\beta)^a c(l_x-e)(l_y-e)}{H_{veg}DV_{veg}\varphi d}\right)^{C2}$ – artificial pine trees $\dfrac{R}{d} = C1 \left(\dfrac{HG_{sb}(cot\beta)^a c^{0.1}(l_x-e)(l_y-e)}{H_{veg}DV_{veg}\varphi d^{0.1}}\right)^{C2}$ – cylindrical model

Reference	Parameter	Equation
Noarayanan et al., (2012a)	Current with tandem arrangement of vegetation (Manning's coefficient)	$n_{Tan} = 0.0422 \times \left(\dfrac{EI*\left(\frac{BG}{D}\right)}{\rho h_u V_{avg}^2 l^3 \left(\frac{SP}{D}\right)}\right)^{0.1332} \times \left(\dfrac{V_{avg}}{\sqrt{gh_{avg}}}\right)^{-0.2217}$ for $SP/D \neq {\sim}3.75$ and $V_r \neq {\sim}4.0$ $n_{Tan} = 0.229 \times \left(\dfrac{EI*(BG/D)}{\rho h_u V_{avg}^2 l^3 (SP/D)}\right)^{0.01812} \times \left(\dfrac{V_{avg}}{\sqrt{gh_{avg}}}\right)^{1.26}$ For $SP/D = {\sim}3.75$ and $V_r = {\sim}4.0$
Noarayanan et al., (2012b)	Current with staggered arrangement of vegetation (Manning's coefficient)	$f_{Stg} = 0.5811 \times \left(\dfrac{EI*\left(\frac{BG}{D}\right)}{\rho h_u V_{avg}^2 l^3 \left(\frac{SP}{D}\right)}\right)^{0.0968} \times \left(\dfrac{V_{avg}}{\sqrt{gh_{avg}}}\right)^{-0.3115}$ for $SP/D \neq {\sim}3.75$ and $V_r \neq {\sim}4.0$ $f_{Stg} = 0.1960 \times \left(\dfrac{EI*(BG/D)}{\rho h_u V_{avg}^2 l^3 (SP/D)}\right)^{1.4136} \times \left(\dfrac{V_{avg}}{\sqrt{gh_{avg}}}\right)^{-3.098}$ for $SP/D = {\sim}3.75$ and $V_r = {\sim}4.0$
Maza et al., (2013)	(Drag coefficient)	$C_d = \left(\dfrac{4600}{Re}\right)^{0.41} + 1.16$
He et al., (2019)	Regular wave (Wave attenuation and Drag coefficient)	$\dfrac{H}{H_0} = -3.92 \times$ $\left(\dfrac{L^2 H_0}{h^3}\right)^{0.18} \left(\dfrac{h_s}{s}\right)^{-44.524} \left[\left(\dfrac{h_r}{h}\right)^{-0.02}\right]^{i} \left[\left(\dfrac{h_c}{h}\right)^{62.40}\right]^{j} \left(\dfrac{B}{L}\right)^{-96.37} (\psi)^{4.528}$ $C_d = 18.025 \exp\left(-0.043 KC\right)$ for $10 \leq KC \leq 37$ $C_d = \dfrac{51.456 \exp\left(-0.02 QC\right)}{QC^{0.457}}$ for $10 \leq Q \leq 52$ $C_d = \dfrac{23.326 \exp\left(-0.601 Wr\right)}{Wr^{0.24}}$ for $0.25 \leq Q \leq 10$
Wang et al., (2022)	Regular waves (Drag coefficient)	$C_d = \left(\dfrac{0.77}{KC_{rv}}\right)^{0.41} + 0.42$
Hari Ram et al., (2022)	Solitary wave (Wave attenuation)	$\dfrac{H}{H_o} = 1 + 0.0059 \left(\dfrac{H_o + d}{d_v}\right)^{-11.21} - 0.1006 \left(\dfrac{Wb}{Wb_{max}}\right)^{0.944} -$ $0.0208 \left(\dfrac{H_o}{d}\right)^{0.1349}$

5. Numerical studies

The shortcomings in the experiment and field studies were explored to be addressed through numerical modelling. Mendez and Losada, [2004] extended the work of Dalrymple *et al.*, [1984] by including the varying depths and the effects of wave damping due to vegetation and wave breaking. The above stated formulation was implemented with a number of assumptions in a full spectrum model of SWAN by Suzuki *et al.*, [2012]. Herein, the formulation ignored the vegetation motion and the effects due to plant flexibility which was included by Asano *et al.*, [1992].

It was observed that the pressure-induced drag was higher than the frictional drag. Tang *et al.*, [2013] investigated the wave damping and run-up reduction due to solitary wave. The formulation of Morison *et al.*, [1950] was implemented in the non-linear shallow water equation in this numerical model through the finite volume approach. Different vegetation height, the density of the belt and the diameter of vegetation were investigated, considering solitary wave. Maza *et al.*, [2015b] investigated the solitary wave evolution over vegetation using Hi FOAM. Direct numerical method and macroscopic analysis using drag coefficient terms were studied and compared. In both cases, advantages and disadvantages concerning computational efficiency and the sensitivity with the coefficients for different arrangements was experienced. However, both methods were still shown to predict the wave passing over the vegetation. Wang *et al.*, [2019] reported the wave damping characteristics of vegetated platforms considering solitary wave similar to Figure 8a. The floating platforms used for wave damping is improved in the presence of vegetation. The transmission coefficient decreased with an increase in steepness and the length of the platform. Erduran and Kutija, [2003] used a Quasi3D model including cantilever beam theory incorporating the effects of deflection in the vegetation due to the wave loads and handled the turbulence using parabolic eddy viscosity approach and combining with the mixing length theories. The deflection of vegetation with respect to its stiffness has been studied.

Cao *et al.*, [2015] discussed the importance of the diffraction effects from vegetation patches and the breaking waves using the extended mild slope equation (EMSE). It was also reported that the energy dissipation by vegetation over high frequency was remarkably high compared to that due to the low-frequency waves. It was compared with the SWAN model and addressed the limitation in the SWAN and the EMSE in the vegetation model study in a large domain simulation. It mainly focused on breaking waves and diffraction with the vegetation model. Li and Xie, [2011] accommodated Euler-Bernoulli law along with large eddy simulation approach and studied flow through flexible vegetation. It was observed that both the vegetation resistance and the vertical Reynolds stress decreased with an increase in the flexibility. The spatial and temporal variation of the deflection of the vegetation resembling the 'Honami' phenomenon in the field were observed through this numerical model. Hu *et al.*, [2015] used the Delft3D model and investigated the storm surge reduction offered by the vegetation. The parameters like stem height, the density of vegetation, storm intensity and speed of storm movement were investigated and observed influence in the storm surge reduction offered by the vegetation belt. Among the parameters, the intensity reduction was observed to be highly influenced by the vegetation stem height.

Divya *et al.*, [2020] reported the effects of wave vegetation interaction using meshless based scheme. The effect of the vegetation model to be a cylinder and square strip was investigated for both emergent and submerged types. It was observed that the shape has little influence in emergent conditions, but in the case of submergence, the cylindrical model was more effective in wave attenuation. The wave attenuation characteristics of vegetation as a single patch and two patches were also studied, and the reduction patterns were reported. Zhang *et al.*, [2013] through 2D numerical model based on the VOF method modelled the wave propagation over vegetation considering different flow by incorporating the effect of drag. Test cases with different conditions with rigid, flexible vegetation for different submergence cases and vegetation over slope were studied and validated. Stoesser *et al.*, [2010] investigated the vortex shedding pattern for staggered arrangement of cylinder in an open channel through large eddy simulation. The vegetation density was found to have greater influence when compared to the Reynolds number. Suppasri *et al.*, [2011] developed a fragility curve based on the methodology proposed by Koshimura and Kayaba, [2008] for the coasts of Thailand. This is a damage assessment methodology combining remote sensing for identification of the different components in the domain (buildings, vegetation, etc.) and numerical technique governing the hydrodynamic parameters of the tsunami. Baba *et al.*, [2008] numerically investigated the Great Barrier Reef's role in protecting Australia's coasts during the 2007 Solomon Islands Tsunami.

By simulating the tsunami event over the actual reef and a model in the absence of reef, the significant role of the reefs in protecting the coasts from the tsunami was explained. Barbier *et al.*, [2013] investigated the effects of wetlands from storm surge events using ADCIRC model simulation for the coasts of Southeast Louisiana. The results showed that the wetlands for the modelled cyclones provided attenuation of the surges.

6. Hybrid protection with vegetation

There are undeniable merits for employing either hard measures or soft measures. Hence a combination of both the hard and soft measures, termed as hybrid solution can accommodate the merits of both measures for coastal protection. Through case studies along the U.S coast, Sutton-Grier *et al.*, [2015] has discussed the strengths and limitation of hard, soft and hybrid measures for coastal protection. Likewise, Nakamura, [2022] through several events and adaptation strategies in Japan and other countries reported the concepts of integrating green infrastructure (soft measures) and grey infrastructure (hard measures) by considering climate change.

BioHaven® floating breakwaters (BFB) is a floating breakwater platform with vegetation. It has been successfully installed in South Louisana marshes. It has been observed that this vegetated floating platform has survived through the 90-MPH-category-1 hurricane and a 3–4 feet storm surge. Wang *et al.*, [2019] investigated the characteristics of these types of platforms by numerically modelling them with the OlaFlow solver through a macroscopic approach. The wave attenuation characteristics of the stem and roots of such vegetated platforms were reported. The roots were observed to perform well compared to the stem. Jayanthi *et al.*, [2020] reported the effectiveness of perforated trapezoidal artificial reefs in the Gulf of Mannar, along the Southeast coast of India. These submerged artificial reefs also facilitate the growth of corals over these modular structures. They help in restoring the shorelines of the Vaan Islands.

Through an experimental investigation, Chim *et al.*, [2022] has shown the advantage of combing hard (revetment armoured with Xbloc[Plus]) and soft (Mangroves) measures and reported an additional overtopping reduction of about 18% to 45% is achieved by combining both. The limitation of revetment berm at high water level was overcome by the submerged root system of mangroves for wave dissipation. Further, addition of an extra layer of armour was claimed to reduce the wave overtopping by 95%–98%.

Submerged artificial reef combines the aspect of wave attenuation by structural units which facilitates in increasing the bio-habitat of the sea with coral growth. Studies were carried out considering such structural units (Arnouil *et al.*, [2008], Buccino *et al.*, [2014]). Deployment of Trapezoidal artificial reef in the Vaan Island restoration has prevented erosion and led to formation of spit formation. An area of 3.15 ha has been increased and furthermore, the artificial reef served in enriching the marine ecosystem in that region as reported by Jayanthi *et al.*, [2020]. Lokesha *et al.*, [2019] through a comprehensive experimental study has shown the performance of these perforated reefs (scaled down model adopted for the restoration of afore mentioned Vaan island) in attenuating regular and random waves. Herein, biological features like the corals over the artificial reef were not considered. However, still, the model has shown good performance in wave attenuation. Thus, in actual field conditions, the natural coral adhering to these artificial submerged reefs would further aid in attenuating extreme waves.

Zaha *et al.*, [2019] considered a combination of coastal protection measures. Coastal dike, moat and vegetation were investigated. It was reported that in hybrid system considering the above coastal protection components, placing emerged vegetation, moat and embankment in an order from the seaward is more effective at the cost of large space in the landward side. In case of insufficient availability of area in the landward side, the order of embankment, moat and vegetation is found to be an optimal solution, where the embankment and moat can be extended seaward.

Kimiwada *et al.*, [2020] also reported the same configuration order of hybrid solution considering submerged vegetation, moat and embankment. The reduction in fluid force index and momentum index on the landward side was higher in the above mentioned hybrid protection. Hybrid protection against tsunami has been implemented in Shichigahama, Japan, wherein, the coastal forest has been planted behind the rehabilitated dyke. Takabatake *et al.*, [2022] studies on casualties' reduction in this region considering the 2011 Tohoku Tsunami showed that the coastal vegetation alone reduced the casualty rate up to 54% and coastal dykes performed to be much better by reducing up to 97%. However, the combined measure has been reported to be more effective.

Reports from van Zelst *et al.*, [2021] evaluated that by combining soft measures like coastal vegetation, mangroves, and marshes fronting levees and have claimed a reduction in the global coastal protection costs. Palikans *et al.*, (2022) have discussed three cases studies on the hybrid coastal protection measures that included the associated challenges in their implementation. The case studies considered are (i) Living Breakwaters in New York Harbour attempts to address the aspects of wave attenuation, beach restoration and preventing shoreline erosion during short terms events like storms as well as in long term considering aspects like climate change, (ii) The Coastal Texas Protection and Restoration Study was focused on problems related to habitat loss, erosion, sea-level rise, and storm surges and (iii) The South Bay Salt Pond Restoration Project in a Francisco Bay aimed in habit restoration, protection form tidal flows and provide public access to connect people with the Bay without disturbing the wildlife.

**(a) Floating vegetation platform South Louisana
(Wang *et al.*, 2019)**

**(b) Living break waters – New York
(Palinkas et al., 2022)**

**(c) Corals over artificial reefs
(Jayanthi 2020)**

Fig. 8. Different types of vegetation based coastal protection features.

7. Contradicts on effectiveness of vegetation

Even though, there are several evidences of vegetation, reefs, etc., protecting the coasts, there were few observations by researchers in which it was not providing the desired protection functionality. Latief and Hadi, [2007] reported both aspects, where the areas of vegetation got destroyed due to the overwhelmed tsunami and areas wherein, the vegetation has effectively protected areas from the tsunami. Fritz *et al.*, [2007] field study reported the devastating impact on the coastal forest of Nusa Kambangan from the 2006 Java tsunami. The inundation depth was observed to be 8 m at the location of the forest. Okal *et al.*, [2010] field investigation on the 2009 Samoa Tsunami near Hikuniu Point, at the north-eastern corner of Niuatoputapu Island, northern Tonga observed the destruction of the entire forest from the tsunami impact. Husrin *et al.*, [2015] through numerical analysis with the vegetation belt cultivated as a measure of protection from the future tsunami in Java islands is ineffective and states the topography plays a vital role in that region. Dilmen *et al.*, [2015] studied the damage incurred on coral reefs in Tutuila Islands, American Samoa, US, due to the 2009 Samoa Tsunami. They tried to model the tsunami in the regions and inferred a high degree of variation in the modelled tsunami in the regions with very high coral damage. Obviously, the roughness in their model had some limitations, which did not include the spatial variation in roughness in the damaged coral reef areas. Looking more profound in the aspect of protection provided by the coral reefs.

From the literature furthermore, an idea on qualitative range on the energy dissipating efficiency for different coastal ecosystem for different sea state is provided in Figure 9. Herein, "X" denotes no energy dissipation and the increasing in the size of the arrow represent slight, moderate and considerable influence respectively.

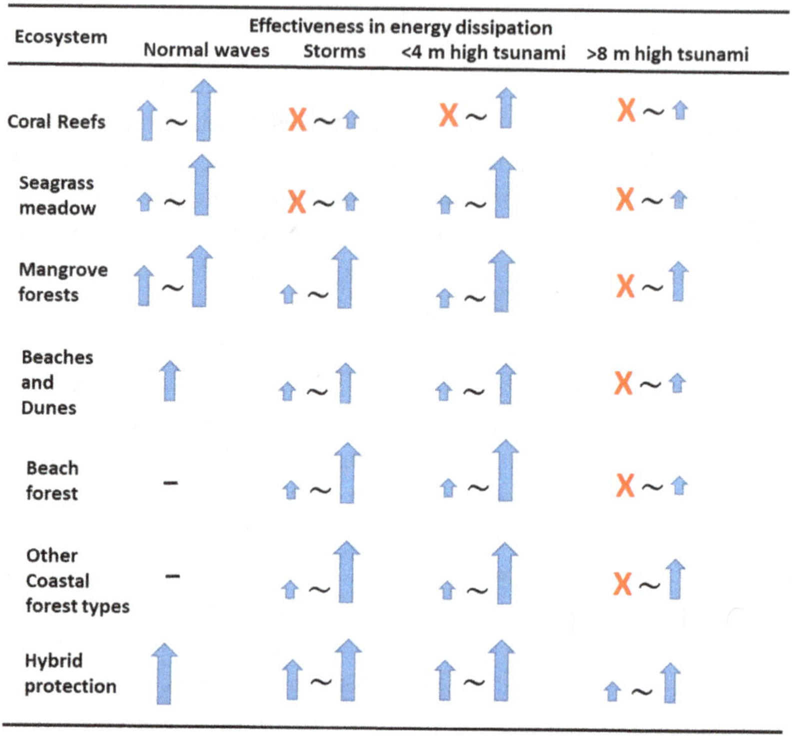

Ecosystem	Effectiveness in energy dissipation			
	Normal waves	Storms	<4 m high tsunami	>8 m high tsunami
Coral Reefs	↑ ~ ↑	X ~ ↑	X ~ ↑	X ~ ↑
Seagrass meadow	↑ ~ ↑	X ~ ↑	↑ ~ ↑	X ~ ↑
Mangrove forests	↑ ~ ↑	↑ ~ ↑	↑ ~ ↑	X ~ ↑
Beaches and Dunes	↑	↑ ~ ↑	↑ ~ ↑	X ~ ↑
Beach forest	–	↑ ~ ↑	↑ ~ ↑	X ~ ↑
Other Coastal forest types	–	↑ ~ ↑	↑ ~ ↑	X ~ ↑
Hybrid protection	↑	↑ ~ ↑	↑ ~ ↑	↑ ~ ↑

Fig. 9. Wave attenuation characteristics for different coastal soft defence components.

8. Summary

The above discussions provide us an insight into the theme of vegetation protecting the coasts from extreme waves. From the collective study on wave vegetation interaction, it is understood that the vegetal characteristics such as the stem diameter, density, arrangement of the vegetation, height of vegetation, and submergence condition of the vegetation during the time of impact are crucial factors in determining the extent of protection that it can provide. In addition, by knowing the characteristics of a tsunami in a particular region from its historical events, one can statistically understand the destructive capacity of the events in that region. The quantity of protection provided by the natural vegetation and the coral reefs also depends on the landscape of the sea bed near the coast and the shoreline at the coast.

There is a lack of complete data on the studies reporting the inefficiency of the vegetation/forest belt in protecting the hinterlands. It is also apparent that the coastal forest fails in the aspects of coastal protection for extreme cases of tsunamis having a high inundation depth. It is also to be noted that an increased inundation depth of a tsunami can be two cases. The tsunamis that are generated have a high magnitude, and the other is the geo-morphological (bathymetry, presence of headlands, confluence coastal regions, etc.). Even though, the experimental and numerical investigations give an insight into the wave-vegetation interactions, all those come with some constraints, especially when considering the interaction of vegetation with tsunami-like waves. In most of the field studies reporting the inefficiency of the vegetation in protecting the coast, there was a lack of data for investigation. The complete set of parameters as listed below for a particular event was not observed in most of the surveys due to the lack of collective knowledge of the wave-vegetation interaction. Thus, the parameters that are vital based on the information gained from the research works on wave-vegetation interaction are as follows

- **Hydrodynamic Parameter**
    - Inundation Height (m)
    - Inundation distance (m)
    - Duration of Impact (hr)
- **Geographic information**
    - Bathymetry of coast
    - Topology of landscape
- **Vegetal Characteristics**
    - Diameter of trunk (m)
    - Height of vegetation (m)
    - Density and arrangement of vegetation
    - Structural strength of the vegetation (EI)
    - Species of vegetation

This provide us the preliminary level of data that must be collected during the survey. The data collection method can be of remote sensing technique, gathering information from local inhabitants, field measured data, etc.

It is undeniable that modelling all the parameters in the field into the experiment or numerical is quite challenging. Some assumptions are made, and some effects may be neglected. For instance, the tsunami evolution during inundation is very challenging. The vegetation will be very small due to the scaling effects. The distorted model study can also be an alternative solution for this problem, yet it is challenging to understand the physics behind the interaction process. The complex geometry of the vegetation, especially its roots and canopy, the complex arrangements. However, data from experiments and numerical studies with such simplified models still help understand many aspects that were not known earlier. The results and observations from the experimental and numerical studies have insight into the parameters that need to be considered while implementing or considering existing vegetation patches. Based on the site location, the extent of protection required, the type of waveform that will be hitting the coast (storms/tsunami), the vegetation density of the vegetation patch, species of vegetation, and length of the patch can be estimated approximately. One of the important aspect that is yet to be explored is the backwash soon after the ingress of the extreme flow like that due to a tsunami which is responsible for the scour under obstructions leading to their stability being questionable. This is apart from the impact of tsunami flow on the motion of debris and floating vessels in the nearshore.

Acknowledgement

This work is supported by the Department of Science & Technology, India Grant No. DST/CCP/CoE/141/2018C under SPLICE – Climate change Programme.

References

Alongi, D.M., 2008. Mangrove forests: resilience, protection from tsunamis, and responses to global climate change. Estuarine, coastal and shelf science, 76(1), pp. 1-13.

Anderson, M.E. and Smith, J.M., 2014. Wave attenuation by flexible, idealized salt marsh vegetation. Coastal Engineering, 83, pp. 82-92.

Arnouil, D.S., 2008. Shoreline response for a Reef Ball™ submerged breakwater system offshore of Grand Cayman Island. Paper of Master Thesis, Florida Institute of Technology.

Asano, T., Deguchi, H. and Kobayashi, N., 1992. Interaction between water waves and vegetation. In Coastal Engineering 1992 (pp. 2709-2723).

Augustin, L.N., Irish, J.L. and Lynett, p. , 2009. Laboratory and numerical studies of wave damping by emergent and near-emergent wetland vegetation. Coastal Engineering, 56(3), pp. 332-340.

Baba, T., Mleczko, R., Burbidge, D., Cummins, p. R. and Thio, H.K., 2009. The effect of the Great Barrier Reef on the propagation of the 2007 Solomon Islands tsunami recorded in northeastern Australia. Tsunami Science Four Years after the 2004 Indian Ocean Tsunami: Part I: Modelling and Hazard Assessment, pp. 2003-2018.

Barbier, E.B., Georgiou, I.Y., Enchelmeyer, B. and Reed, D.J., 2013. The value of wetlands in protecting southeast Louisiana from hurricane storm surges. PloS onc, 8(3), p. c58715.

Besset, M., Gratiot, N., Anthony, E.J., Bouchette, F., Goichot, M. and Marchesiello, p. , 2019. Mangroves and shoreline erosion in the Mekong River delta, Viet Nam. Estuarine, Coastal and Shelf Science, 226, p. 106263.

Bianchi, T.S., Allison, M.A., Zhao, J., Li, X., Comeaux, R.S., Feagin, R.A. and Kulawardhana, R.W., 2013. Historical reconstruction of mangrove expansion in the Gulf of Mexico: linking climate change with carbon sequestration in coastal wetlands. Estuarine, Coastal and Shelf Science, 119, pp. 7-16.

Breithaupt, J.L., Smoak, J.M., Smith III, T.J., Sanders, C.J. and Hoare, A., 2012. Organic carbon burial rates in mangrove sediments: Strengthening the global budget. Global Biogeochemical Cycles, 26(3).

Buccino, M., Del Vita, I. and Calabrese, M., 2014. Engineering modeling of wave transmission of reef balls. Journal of waterway, port, coastal, and ocean engineering, 140(4), p. 04014010.

Busari, A., Gbadebo, O. and Jimoh, I., 2013. Effect Of Vegetation Density On The Hydrodynamic Of Submerged Vegetated Flow.

Cao, H., Feng, W., Hu, Z., Suzuki, T. and Stive, M.J., 2015. Numerical modeling of vegetation-induced dissipation using an extended mild-slope equation. Ocean Engineering, 110, pp. 258-269.

Chang, C.W. and Mori, N., 2021. Green infrastructure for the reduction of coastal disasters: A review of the protective role of coastal forests against tsunami, storm surge, and wind waves. Coastal Engineering Journal, 63(3), pp. 370-385.

Chang, S.E., Adams, B.J., Alder, J., Berke, p. R., Chuenpagdee, R., Ghosh, S. and Wabnitz, C., 2006. Coastal ecosystems and tsunami protection after the December 2004 Indian Ocean tsunami. Earthquake Spectra, 22(3_suppl), pp. 863-887.

Chim, Y.Q., Muilwijk, M., Ruwiel, T., Reedijk, J.S. and Yang, Z.Q., Physical Model Investigation of Hybrid Coastal Protection, combining Nature-Based and Hard Infrastructure.

Cochard, R., 2011. The 2004 tsunami in Aceh and Southern Thailand: Coastal ecosystem services, damages and resilience. The Tsunami Threatâ Research and Technology, pp. 179-216.

Dalrymple, R.A., Kirby, J.T. and Hwang, p. A., 1984. Wave diffraction due to areas of energy dissipation. Journal of waterway, port, coastal, and ocean engineering, 110(1), pp. 67-79.

Danielsen, F., Sørensen, M.K., Olwig, M.F., Selvam, V., Parish, F., Burgess, N.D., Hiraishi, T., Karunagaran, V.M., Rasmussen, M.S., Hansen, L.B. and Quarto, A., 2005. The Asian tsunami: A protective role for coastal vegetation. Science, 310(5748), pp. 643-643.

Das, S. and Vincent, J.R., 2009. Mangroves protected villages and reduced death toll during Indian super cyclone. Proceedings of the National Academy of Sciences, 106(18), pp. 7357-7360.

Dilmen, D.I., Titov, V.V. and Roe, G.H., 2015. Evaluation of the relationship between coral damage and tsunami dynamics; Case study: 2009 Samoa tsunami. Pure and Applied Geophysics, 172, pp. 3557-3572.

Divya, R., Sriram, V. and Murali, K., 2020. Wave-vegetation interaction using improved meshless local Petrov Galerkin method. Applied Ocean Research, 101, p. 102116.

Dogan, G.G., Yalciner, A.C., Yuksel, Y., Ulutaş, E., Polat, O., Güler, I., Şahin, C., Tarih, A. and Kânoğlu, U., 2021. The 30 October 2020 Aegean Sea tsunami: Post-event field survey along Turkish coast. Pure and applied geophysics, 178, pp. 785-812.

Duarte, C.M., Losada, I.J., Hendriks, I.E., Mazarrasa, I. and Marbà, N., 2013. The role of coastal plant communities for climate change mitigation and adaptation. Nature climate change, 3(11), pp. 961-968.

Erduran, K.S. and Kutija, V., 2003. Quasi-three-dimensional numerical model for flow through flexible, rigid, submerged and non-submerged vegetation. Journal of Hydroinformatics, 5(3), pp. 189-202.

Fathi-Moghadam, M., Davoudi, L. and Motamedi-Nezhad, A., 2018. Modeling of solitary breaking wave force absorption by coastal trees. Ocean Engineering, 169, pp. 87-98.

Fathi-Moghadam, M., Kashefipour, M., Ebrahimi, N. and Emamgholizadeh, S., 2011. Physical and numerical modeling of submerged vegetation roughness in rivers and flood plains. Journal of Hydrologic Engineering, 16(11), pp. 858-864.

Feagin, R.A., Mukherjee, N., Shanker, K., Baird, A.H., Cinner, J., Kerr, A.M., Koedam, N., Sridhar, A., Arthur, R., Jayatissa, L.P. and Lo Seen, D., 2010. Shelter from the storm? Use and misuse of coastal vegetation bioshields for managing natural disasters. Conservation Letters, 3(1), pp. 1-11.

Fischenich, J.C., 2000. Resistance due to vegetation.

Freeman, G.E., Rahmeyer, W.H. and Copeland, R.R., 2000. Determination of resistance due to shrubs and woody vegetation.

Fritz, H.M., Kongko, W., Moore, A., McAdoo, B., Goff, J., Harbitz, C., Uslu, B., Kalligeris, N., Suteja, D., Kalsum, K. and Titov, V., 2007. Extreme runup from the 17 July 2006 Java tsunami. Geophysical Research Letters, 34(12).

Fritz, H.M., Petroff, C.M., Catalán, P.A., Cienfuegos, R., Winckler, P., Kalligeris, N., Weiss, R., Barrientos, S.E., Meneses, G., Valderas-Bermejo, C. and Ebeling, C., 2011. Field survey of the 27 February 2010 Chile tsunami. Pure and Applied Geophysics, 168, pp. 1989-2010.

Gedan, K.B., Kirwan, M.L., Wolanski, E., Barbier, E.B. and Silliman, B.R., 2011. The present and future role of coastal wetland vegetation in protecting shorelines: answering recent challenges to the paradigm. Climatic change, 106, pp. 7-29.

Gusiakov, V.K., 2009. Tsunami history: Recorded. The sea, 15, pp. 23-53.

Han, P. and Yu, X., 2022. An unconventional tsunami: 2022 Tonga event. Physics of Fluids, 34(11).

Hari Ram, N., Sriram, V. and Murali, K., 2022. Experimental investigation on the characteristics of solitary and elongated solitary waves passing over vegetation belt. Journal of Ocean Engineering and Marine Energy, 8(3), pp. 305-318.

Harris, L.E., 2009. Artificial reefs for ecosystem restoration and coastal erosion protection with aquaculture and recreational amenities. Reef Journal, 1(1), pp. 235-246.

He, F., Chen, J. and Jiang, C., 2019. Surface wave attenuation by vegetation with the stem, root and canopy. Coastal Engineering, 152, p. 103509.

Hiraishi, T., 2003. Greenbelt tsunami prevention in South-Pacific region. Report of the Port and Airport Research Institute, 42(2), pp. 1-23.

Hu, K., Chen, Q. and Wang, H., 2015. A numerical study of vegetation impact on reducing storm surge by wetlands in a semi-enclosed estuary. Coastal Engineering, 95, pp. 66-76.

Hu, Z., Lian, S., Zitman, T., Wang, H., He, Z., Wei, H., Ren, L., Uijttewaal, W. and Suzuki, T., 2022. Wave breaking induced by opposing currents in submerged vegetation canopies. Water Resources Research, 58(4), p. e2021WR031121.

Huang, Z., Yao, Y., Sim, S.Y. and Yao, Y., 2011. Interaction of solitary waves with emergent, rigid vegetation. Ocean Engineering, 38(10), pp. 1080-1088.

Husrin, S., Kelvin, J., Putra, A., Prihantono, J., Cara, Y. and Hani, A., 2015. Assessment on the characteristics and the damping performance of coastal forests in Pangandaran after the 2006 Java Tsunami. Procedia Earth and Planetary Science, 12, pp. 20-30.

Irish, J.L., Ewing, L.C. and Jones, C.P., 2012. Observations from the 2009 Samoa tsunami: damage potential in coastal communities. Journal of waterway, port, coastal, and ocean engineering, 138(2), pp. 131-141.

Irtem, E., Gedik, N., Kabdasli, M.S. and Yasa, N.E., 2009. Coastal forest effects on tsunami run-up heights. Ocean Engineering, 36(3-4), pp. 313-320.

Ishikawa, Y., Mizuhara, K. and Ashida, S., 2000. Effect of density of trees on drag exerted on trees in river channels. Journal of Forest Research, 5, pp. 271-279.

Ismail, H., Abd Wahab, A.K. and Alias, N.E., 2012. Determination of mangrove forest performance in reducing tsunami run-up using physical models. Natural hazards, 63, pp. 939-963.

Jayanthi, M., Patterson Edward, J.K., Malleshappa, H., Gladwin Gnana Asir, N., Mathews, G., Diraviya Raj, K., Bilgi, D.S., Ashok Kumar, T.K. and Sannasiraj, S.A., 2020. Perforated trapezoidal artificial reefs can augment the benefits of restoration of an island and its marine ecosystem. Restoration Ecology, 28(1), pp. 233-243.

Kimiwada, Y., Tanaka, N. and Zaha, T. (2020). Differences in effectiveness of a hybrid tsunami defense system comprising an embankment, moat, and forest in submerged, emergent, or combined conditions. Ocean Engineering, 208, 107457.

Koshimura, S. and Kayaba, S., 2008, September. Tsunami damage detection using high-resolution optical satellite imagery. In Proceedings of the 6th International Workshop on Remote Sensing for Post Disaster Response, University of Pavia, Italy (pp. 11-12).

Krauss, K.W., Doyle, T.W., Doyle, T.J., Swarzenski, C.M., From, A.S., Day, R.H. and Conner, W.H., 2009. Water level observations in mangrove swamps during two hurricanes in Florida. Wetlands, 29, pp. 142-149.

Lavigne, F., Gomez, C., Giffo, M., Wassmer, p. , Hoebreck, C., Mardiatno, D., Prioyono, J. and Paris, R., 2007. Field observations of the 17 July 2006 Tsunami in Java. Natural Hazards and Earth System Sciences, 7(1), pp. 177-183.

Lakshmanan, N., Kantharaj, M. and Sundar, V., 2012. The effects of flexible vegetation on forces with a Keulegan-Carpenter number in relation to structures due to long waves. Journal of Marine Science and Application, 11, pp. 24-33.

Latief, H. and Hadi, S., 2007. The role of forests and trees in protecting coastal areas against tsunamis. In Regional Technical Workshop on Coastal Protection in the Aftermath of the Indian Ocean Tsunami What Role for Forests Trees (pp. 5-32).

Lei, J. and Nepf, H., 2019. Wave damping by flexible vegetation: Connecting individual blade dynamics to the meadow scale. Coastal Engineering, 147, pp. 138-148.

Li, C.W. and Xie, J.F., 2011. Numerical modeling of free surface flow over submerged and highly flexible vegetation. Advances in Water Resources, 34(4), pp. 468-477.

Lokesha, Sannasiraj, S.A. and Sundar, V., 2019. Hydrodynamic characteristics of a submerged trapezoidal artificial reef unit. Proceedings of the Institution of Mechanical Engineers, Part M: Journal of Engineering for the Maritime Environment, 233(4), pp. 1226-1239.

Lou, S., Chen, M., Ma, G., Liu, S. and Zhong, G., 2018. Laboratory study of the effect of vertically varying vegetation density on waves, currents and wave-current interactions. Applied Ocean Research, 79, pp. 74-87.

Marois, D.E. and Mitsch, W.J., 2015. Coastal protection from tsunamis and cyclones provided by mangrove wetlands–a review. International Journal of Biodiversity Science, Ecosystem Services & Management, 11(1), pp. 71-83.

Mascarenhas, A. and Jayakumar, S., 2008. An environmental perspective of the post-tsunami scenario along the coast of Tamil Nadu, India: Role of sand dunes and forests. Journal of Environmental Management, 89(1), pp. 24-34.

Maza, M., Lara, J.L. and Losada, I.J., 2013. A coupled model of submerged vegetation under oscillatory flow using Navier–Stokes equations. Coastal Engineering, 80, pp. 16-34.

Maza, M., Lara, J.L., Losada, I.J., Ondiviela, B., Trinogga, J. and Bouma, T.J., 2015a. Large-scale 3-D experiments of wave and current interaction with real vegetation. Part 2: Experimental analysis. Coastal Engineering, 106, pp. 73-86.

Maza, M., Lara, J.L. and Losada, I.J., 2015b. Tsunami wave interaction with mangrove forests: A 3-D numerical approach. Coastal Engineering, 98, pp. 33-54.

Maza, M., Lara, J.L. and Losada, I.J., 2016. Solitary wave attenuation by vegetation patches. Advances in Water Resources, 98, pp. 159-172.

Meijer, D.G. and Van Velzen, E.H., 1999, August. Prototype-scale flume experiments on hydraulic roughness of submerged vegetation. In 28th International IAHR Conference, Graz.

Mendez, F.J. and Losada, I.J., 2004. An empirical model to estimate the propagation of random breaking and nonbreaking waves over vegetation fields. Coastal Engineering, 51(2), pp. 103-118.

Mikami, T., Shibayama, T., Esteban, M., Takabatake, T., Nakamura, R., Nishida, Y., Achiari, H., Rusli, Marzuki, A.G., Marzuki, M.F.H. and Stolle, J., 2019. Field survey of the 2018 Sulawesi tsunami: Inundation and run-up heights and damage to coastal communities. Pure and Applied Geophysics, 176, pp. 3291-3304.

Mimura, N., Yasuhara, K., Kawagoe, S., Yokoki, H. and Kazama, S., 2011. Damage from the Great East Japan Earthquake and Tsunami-a quick report. Mitigation and adaptation strategies for global change, 16, pp. 803-818.

Montakhab, A., Yusuf, B., Ghazali, A.H. and Mohamed, T.A., 2012. Flow and sediment transport in vegetated waterways: A review. Reviews in Environmental Science and Bio/Technology, 11, pp. 275-287.

Morison, J.R., Johnson, J.W. and Schaaf, S.A., 1950. The force exerted by surface waves on piles. Journal of Petroleum Technology, 2(05), pp. 149-154.

Muhari, A., Imamura, F., Arikawa, T., Hakim, A.R. and Afriyanto, B., 2018. Solving the puzzle of the September 2018 Palu, Indonesia, tsunami mystery: Clues from the tsunami waveform and the initial field survey data. Journal of Disaster Research, 13(Scientific Communication), p. sc20181108.

Nakamura, F., 2022. Green Infrastructure and Climate Change Adaptation: Function, Implementation and Governance (p. 506). Springer Nature.

Noarayanan, L., Murali, K. and Sundar, V., 2012a. Manning's 'n' co-efficient for flexible emergent vegetation in tandem configuration. Journal of hydro-environment research, 6(1), pp. 51-62.

Noarayanan, L., Murali, K. and Sundar, V., 2012b. Performance of flexible emergent vegetation in staggered configuration as a mitigation measure for extreme coastal disasters. Natural hazards, 62, pp. 531-550.

Noarayanan, L., Murali, K. and Sundar, V., 2012c. Role of vegetation on beach run-up due to regular and cnoidal waves. Journal of Coastal Research, 28(1A), pp. 123-130.

Okal, E.A., Fritz, H.M., Synolakis, C.E., Borrero, J.C., Weiss, R., Lynett, P.J., Titov, V.V., Foteinis, S., Jaffe, B.E., Liu, P.L.F. and Chan, I.C., 2010. Field survey of the Samoa tsunami of 29 September 2009. Seismological Research Letters, 81(4), pp. 577-591.

Palermo, D., Nistor, I., Saatcioglu, M. and Ghobarah, A., 2013. Impact and damage to structures during the 27 February 2010 Chile tsunami. Canadian Journal of Civil Engineering, 40(8), pp. 750-758.

Panigrahi, K. and Khatua, K.K., 2015. Prediction of velocity distribution in straight channel with rigid vegetation. Aquatic Procedia, 4, pp. 819-825.

Righetti, M., 2008. Flow analysis in a channel with flexible vegetation using double-averaging method. Acta Geophysica, 56, pp. 801-823.

Rodriguez, H., Wachtendorf, T., Kendra, J. and Trainor, J., 2006. A snapshot of the 2004 Indian Ocean tsunami: societal impacts and consequences. Disaster Prevention and Management: An International Journal, 15(1), pp. 163-177.

Satake, K., Nishimura, Y., Putra, P.S., Gusman, A.R., Sunendar, H., Fujii, Y., Tanioka, Y., Latief, H. and Yulianto, E., 2013. Tsunami source of the 2010 Mentawai, Indonesia earthquake inferred from tsunami field survey and waveform modeling. Pure and Applied Geophysics, 170, pp. 1567-1582.

Sheth, A., Sanyal, S., Jaiswal, A. and Gandhi, p. , 2006. Effects of the December 2004 Indian Ocean tsunami on the Indian mainland. Earthquake spectra, 22(3_suppl), pp. 435-473.

Shuto, N., 1987. The effectiveness and limit of tsunami control forests. Coastal Engineering in Japan, 30(1), pp. 143-153.

Stoesser, T., Kim, S.J. and Diplas, p. , 2010. Turbulent flow through idealized emergent vegetation. Journal of Hydraulic Engineering, 136(12), pp. 1003-1017.

Struve, J., Falconer, R. and Wu, Y., 2003. Influence of model mangrove trees on the hydrodynamics in a flume. Estuarine, Coastal and Shelf Science, 58(1), pp. 163-171.

Sundar, V., Murali, K. and Noarayanan, L., 2011. Effect of vegetation on run-up and wall pressures due to cnoidal waves. Journal of hydraulic research, 49(4), pp. 562-567.

Sundar, V., Sannasiraj, S.A., Murali, K. and Sundaravadivelu, R., 2007. Runup and inundation along the Indian peninsula, including the Andaman Islands, due to Great Indian Ocean Tsunami. Journal of waterway, port, coastal, and ocean engineering, 133(6), pp. 401-413.

Sundar, V., Sannasiraj, S.A., Murali, K. and Sriram, V., 2020. Tsunami: Engineering Perspective For Mitigation, Protection And Modeling (Vol. 50). World Scientific.

Suppasri, A., Koshimura, S. and Imamura, F., 2011. Developing tsunami fragility curves based on the satellite remote sensing and the numerical modeling of the 2004 Indian Ocean tsunami in Thailand. Natural Hazards and Earth System Sciences, 11(1), pp. 173-189.

Sutton-Grier, A.E., Wowk, K. and Bamford, H., 2015. Future of our coasts: The potential for natural and hybrid infrastructure to enhance the resilience of our coastal communities, economies and ecosystems. Environmental Science & Policy, 51, pp. 137-148.

Suzuki, T., Zijlema, M., Burger, B., Meijer, M.C. and Narayan, S., 2012. Wave dissipation by vegetation with layer schematization in SWAN. Coastal Engineering, 59(1), pp. 64-71.

Takabatake, T., Esteban, M. and Shibayama, T., 2022. Simulated effectiveness of coastal forests on reduction in loss of lives from a tsunami. International Journal of Disaster Risk Reduction, 74, p. 102954.

Tanaka, N., 2009. Vegetation bioshields for tsunami mitigation: review of effectiveness, limitations, construction, and sustainable management. Landscape and Ecological Engineering, 5, pp. 71-79.

Tanaka, N., Sasaki, Y., Mowjood, M.I.M., Jinadasa, K.B.S.N. and Homchuen, S., 2007. Coastal vegetation structures and their functions in tsunami protection: experience of the recent Indian Ocean tsunami. Landscape and ecological engineering, 3, pp. 33-45.

Tang, J., Causon, D., Mingham, C. and Qian, L., 2013. Numerical study of vegetation damping effects on solitary wave run-up using the nonlinear shallow water equations. Coastal Engineering, 75, pp. 21-28.

Tang, J., Shen, Y., Causon, D.M., Qian, L. and Mingham, C.G., 2017. Numerical study of periodic long wave run-up on a rigid vegetation sloping beach. Coastal Engineering, 121, pp. 158-166.

Thampanya, U., Vermaat, J.E., Sinsakul, S. and Panapitukkul, N., 2006. Coastal erosion and mangrove progradation of Southern Thailand. Estuarine, coastal and shelf science, 68(1-2), pp. 75-85.

Tinti, S., Maramai, A., Armigliato, A., Graziani, L., Manucci, A., Pagnoni, G. and Zaniboni, F., 2006. Observations of physical effects from tsunamis of December 30, 2002 at Stromboli volcano, southern Italy. Bulletin of volcanology, 68, pp. 450-461.

Tognin, D., Peruzzo, P., De Serio, F., Ben Meftah, M., Carniello, L., Defina, A. and Mossa, M., 2019. Experimental setup and measuring system to study solitary wave interaction with rigid emergent vegetation. Sensors, 19(8), p. 1787.

van Zelst, V.T., Dijkstra, J.T., van Wesenbeeck, B.K., Eilander, D., Morris, E.P., Winsemius, H.C., Ward, p. J. and de Vries, M.B., 2021. Cutting the costs of coastal protection by integrating vegetation in flood defences. Nature communications, 12(1), p. 6533.

Wang, H., Yin, Z., Luan, Y., Wang, Y. and Liu, D., 2022a. Hydrodynamic characteristics of idealized flexible vegetation under regular waves: Experimental investigations and analysis. Journal of Coastal Research, 38(3), pp. 673-680.

Wang, Y., Yin, Z. and Liu, Y., 2019. Numerical study of solitary wave interaction with a vegetated platform. Ocean Engineering, 192, p. 106561.

Wang, Y., Yin, Z. and Liu, Y., 2022b. Experimental investigation of wave attenuation and bulk drag coefficient in mangrove forest with complex root morphology. Applied Ocean Research, 118, p. 102974.

Ward, R.D., 2020. Carbon sequestration and storage in Norwegian Arctic coastal wetlands: Impacts of climate change. Science of the Total Environment, 748, p. 141343.

Wu, F.C., Shen, H.W. and Chou, Y.J., 1999. Variation of roughness coefficients for unsubmerged and submerged vegetation. Journal of Hydraulic Engineering, 125(9), pp. 934-942.

Wu, W.C. and Cox, D.T., 2016. Effects of vertical variation in vegetation density on wave attenuation. Journal of Waterway, Port, Coastal, and Ocean Engineering, 142(2), p. 04015020.

Zaha, T., Tanaka, N. and Kimiwada, Y., 2019. Flume experiments on optimal arrangement of hybrid defense system comprising an embankment, moat, and emergent vegetation to mitigate inundating tsunami current. Ocean Engineering, 173, pp. 45-57.

Zhang, M., Hao, Z., Zhang, Y. and Wu, W., 2013. Numerical simulation of solitary and random wave propagation through vegetation based on VOF method. Acta Oceanologica Sinica, 32, pp. 38-46.

Zhang, Y., Tang, C. and Nepf, H., 2018. Turbulent kinetic energy in submerged model canopies under oscillatory flow. Water Resources Research, 54(3), pp. 1734-1750.

Zhao, C., Tang, J. and Shen, Y., 2021. Experimental study on solitary wave attenuation by emerged vegetation in currents. Ocean Engineering, 220, p. 108414.

Zhao, M. and Fan, Z., 2017. Hydrodynamic characteristics of submerged vegetation flow with non-constant vertical porosity. Plos one, 12(4), p. e0176712.

Index

A

aboveground biomass, 12
aquatic species, 4
attenuation processes, 8
belowground biomass, 11, 18
biomass, 76
biomechanical properties, 59, 63, 82
biomechanics, 14
bottom friction, 71

B

Boussinesq-type equations, 156
Building with Nature, 2
buoyancy force, 126

C

cantilever beam models, 122
Cauchy number, 121, 127
Chorin time split algorithm, 166
circular cylinders, 184
coastal forest, 29, 41, 92, 97, 98, 108, 117
coastal protection, 55, 56, 57, 222
coastal vegetation, 207
coefficient of drag, 173
COULWAVE, 159
coupling, 5
cylindrical vegetation, 171

D

damage mitigation, 30
damping coefficient, 133
damping, 5
Darcian velocity, 164
Darcy law, 148
density, 39
density of the vegetation, 37
diameter of vegetation, 220
Direct Numerical Simulation, 185
drag coefficients, 151
drag force, 62, 212
drag resistance, 150

E

ecosystems, 2, 4
eddy-resolving, 5
emergent condition, 215
emergent mangrove forests, 2
emergent vegetation, 4, 102, 217
energy attenuation, 66, 69, 78
energy dissipation, 40, 146, 213, 225
energy reduction, 29, 40, 95
Eulerian framework, 160
excursion ratio, 121
extreme waves, 4

F

FEBOUSS, 157